AF766726

EPISTEMIC JUSTICE, MINDFULNESS, AND THE ENVIRONMENTAL HUMANITIES

Epistemic Justice, Mindfulness, and the Environmental Humanities explores how contemplative pedagogies and mindfulness can be used in the classroom to address epistemic and environmental injustice.

In recent years, there has been a groundswell of interest in contemplative pedagogies in higher education, with increasing attention from the environmental sciences, environmental humanities, and sustainability studies. Teachers and writers have demonstrated how mindfulness practices can be a key to anti-oppression and anti-racist efforts, both in and out of the classroom. Not all forms of contemplative pedagogy are suited for this anti-colonial and anti-oppressive resistance, however. Simply adopting mindfulness practices in the classroom is not enough to dislodge and dismantle white supremacy in higher education. *Epistemic Justice, Mindfulness, and the Environmental Humanities* advocates for mindfulness practices that affirm multiple epistemologies and cultural traditions. Written for educators in the environmental humanities and other related disciplines, the chapters interrogate the western uptake of mindfulness practices and suggest anti-colonial and anti-oppressive methods for bringing mindfulness into the classroom. The chapters also discuss what mindfulness practices have to offer to the pursuit of a culturally relevant pedagogy.

This highly applied and practical text will be an insightful read for educators in the environmental humanities and across the liberal arts disciplines.

Janelle Adsit is Associate Professor of English at Humboldt State University, USA.

EPISTEMIC JUSTICE, MINDFULNESS, AND THE ENVIRONMENTAL HUMANITIES

Reflections on Teaching

Janelle Adsit

Routledge
Taylor & Francis Group
LONDON AND NEW YORK

First published 2022
by Routledge
2 Park Square, Milton Park, Abingdon, Oxon OX14 4RN

and by Routledge
605 Third Avenue, New York, NY 10158

Routledge is an imprint of the Taylor & Francis Group, an informa business

© 2022 Janelle Adsit

The right of Janelle Adsit to be identified as author of this work has been asserted by her in accordance with sections 77 and 78 of the Copyright, Designs and Patents Act 1988.

All rights reserved. No part of this book may be reprinted or reproduced or utilised in any form or by any electronic, mechanical, or other means, now known or hereafter invented, including photocopying and recording, or in any information storage or retrieval system, without permission in writing from the publishers.

A part of chapter 4 "Reading and writing practices in the classroom" was previously published in Janelle Adsit and Renée Byrd, Writing Intersectional Identities (Bloomsbury Academic, 2019).

Trademark notice: Product or corporate names may be trademarks or registered trademarks, and are used only for identification and explanation without intent to infringe.

British Library Cataloguing-in-Publication Data
A catalogue record for this book is available from the British Library

Library of Congress Cataloging-in-Publication Data
Names: Adsit, Janelle, author.
Title: Epistemic justice, mindfulness, and the environmental humanities : reflections on teaching / Janelle Adsit.
Description: Abingdon, Oxon ; New York, NY : Routledge, 2022. | Includes bibliographical references and index.
Identifiers: LCCN 2021025094 (print) | LCCN 2021025095 (ebook) | ISBN 9780367479633 (hardback) | ISBN 9780367479626 (paperback) | ISBN 9781003037439 (ebook)
Subjects: LCSH: Human ecology and the humanities. | Environmental justice. | Mindfulness (Psychology) | Culturally relevant pedagogy.
Classification: LCC GF22 .A47 2022 (print) | LCC GF22 (ebook) | DDC 304.2—dc23
LC record available at https://lccn.loc.gov/2021025094
LC ebook record available at https://lccn.loc.gov/2021025095

ISBN: 978-0-367-47963-3 (hbk)
ISBN: 978-0-367-47962-6 (pbk)
ISBN: 978-1-003-03743-9 (ebk)

DOI: 10.4324/9781003037439

Typeset in Bembo
by Apex CoVantage, LLC

CONTENTS

ACKNOWLEDGMENTS

Each page of this book is written in deep acknowledgment of the privilege of having time, resources, and space to write and, most consequentially for me, in being able to listen and receive. This book emerged through many conversations, webinars, seminars, and because of the time I had to read and learn from scholars through their pages. These words were written in a period of time when over four million people lost their lives to COVID-19 and millions more put themselves at risk by serving their communities day after day. A disproportionate number of these are Black and Indigenous peoples and people of color. I write every page of this book in acknowledgment of their lives.

I want to thank those who have shown up on Zoom calls and in socially distanced outdoor spaces to talk with me. Thank you to these voices I trust to help me see beyond some of the knowledges I've received.

Thanks to John Johnson who has shaped my thinking. Thank you for caring that this book would come into existence and for the years we shared.

My never-ending appreciation also goes to Sue Doe for her mentorship and collaboration over many years. Abiding gratitude to Conchitina Cruz for friendship, for modeling what it means to do meaningful work.

Deep gratitude to Shawn Hargan also: for words that keep me going—thank you for always being family to me.

Thanks to Greta Gaard and Bengu Erguner-Tekinalp for sharing resources and expertise and offering bright news and meaningful collaboration in a difficult time.

Thank you to Reverend Deborah Johnson for your teaching and your vision.

Resounding thanks to the students who have allowed me to accompany them in journeys of thought that I have learned much from: Natalie Acuña, Megan Awwad, Evie Ferreira, Kelly Fortner, Lucy Gutierrez, Alysia Hegg, Meriel Melendrez Mees, Dakota Rohlin, Em Scott, Lex Siebuhr, Vanessa Tenorio, and Madi Whaley. My humble gratitude goes also to Cristian Martinez, Erika Andrews,

Theressa Lopez, Marcos Hernandez, Hallie Lepphaille, and Corrina Wells for allowing me to be your colleague and collaborator. My sincere thanks to other colleagues at Humboldt State, especially Renée Byrd and Sarah Jaquette Ray for their companionship and care and for the inspiration they share with me. Thanks also to Christina Hsu Accomando, Sarah Ben-Zvi, Deepti Chatti, Andrea Delgado, Laura Hahn, Heal McKnight, Carly Marino, Edelmira Reynoso, Erin Sullivan, Janet Winston, Andrea Juarez, Kirby Moss, and Patty Yancey. Big-hearted thanks to Lisa Tremain for partnership and to Mary Ann Creadon for her vote of confidence. Much appreciation to Teri Bronder-Lewis for her strength and wisdom.

This book is informed by instruction and resources shared by Cutcha Risling Baldy, Kaitlin Reed, and Kayla Begay; thank you for all you've taught me in each space we've shared. This book was written before the start of the Spring 2021 book group co-facilitated by Professors Risling Baldy and Reed, but I want to acknowledge the learning I know I will do there. I also wish to thank additional faculty from the Environmental Studies and Environment & Community programs who have informed my thinking. Thanks also to the Critical Race, Gender, and Sexuality Studies faculty for allowing me to join you and learn from you.

Thank you to the HSU library, especially Brianne Hagen who made herself available to track down many sources for this book. Thank you for your meta-level wondrousness, Bri.

Kumi Watanabe-Schock and Jay Schock have my endless admiration and gratitude; I can't thank you both enough. You make things happen and you make things possible. Thank you for your sustaining friendship.

Thank you to Grace Harrison, Rosie Anderson, and Rebecca Brennan at Routledge for their support, patience, and guidance as we completed this book in a tumultuous year.

To my parents, Eric and Lisa Adsit, who are behind everything I've ever done. Thank you for all the ways that you daily encourage me forward. You set me in motion and keep me going. Thank you.

I wrote this book on the present and ancestral homeland and unceded territory of the Wiyot Tribe, near to the Hupa, Karuk, Mattole, Tolowa, Wailaki, and Yurok Tribal communities. Specifically, this book was written in Jaroujiji, which, in my limited understanding, roughly translates as "where you sit and rest." The Redwood trees, the Bay, and the roll of the ocean make this a meditative place. I make this acknowledgment in recognition of the remarkable fortune it is to be able to sit and rest in this world—what should be an entitlement of all is denied to so many.

Reverend angel Kyodo williams[1] calls us to account for "Our inability . . . to honor the theft of these lands and the building of wealth, power, and privilege on the countless backs and graves of Black people." I make this acknowledgement in accountability to this reality with the intention to do all I can to redress what has been done and halt these systems of oppression. May my words match my actions.

Janelle Adsit
January 2021

Note

1 williams, angel Kyodo, Lama Rod Owens, and Jasmine Syedullah, *Radical Dharma: Talking Race, Love, and Liberation* (North Atlantic, 2016).

INTRODUCTION

"Teachers, We Cannot Go Back to the Way Things Were."[1] This headline from Bettina Love's April 2020 *EdWeek* article has stayed with me from the day it was published, as I have tried to imagine my part in higher education in this pandemic-changed and climate-changed world. The way things were—and are—can be described with words like marginalization, disposability, massive student debt, mass incarceration, police brutality, border patrol, toxins distributed by racist paradigms, and precarity that is unevenly experienced. These words are tools for naming the daily experience of living in oppression and having to fight to matter in this world. Things never should have been the way they were, Bettina Love notes, and we cannot continue.

Bettina Love's work is abolitionist. She calls us to understand the interlocking systems that make state violence possible and that perpetuate state violence in our schools. The struggle against these forces—and the racism, sexism, homophobia, transphobia, Islamophobia, ableism, and classism that pervades—must be at the forefront of our work as teachers, as researchers, and as activists. The call comes clearly: we need to transform. The transformation is long overdue.

Education should have never been this way, serving some to the exclusion of many more, keeping systems of privilege in place. And what is said of education broadly here also pertains to our fields of study in the environmental humanities. The construction of the humanities within western(ized) systems of higher education was built upon and made to reinforce the hierarchies that center and affirm the lives of some while making "others" disposable.

What does it mean to facilitate a classroom space—whether that space is virtual or within four walls—in this time, an era that is characterized by the long and continued legacy of colonial and racist thinking that shapes the conditions of the living beings of this planet? What does it mean to facilitate spaces populated by people who are variously benefitting from or brutalized by this system? We call these

DOI: 10.4324/9781003037439-1

spaces environments of learning, but they are often experienced as sites of trauma, exclusion, and subjugation.

This book is about the educational system that the environmental humanities is part of. The environmental humanities is a teaching discipline as much as it is a research discipline, and our commitments to environmental justice should span across the realms of our work. And of course, these realms are not separate. As Natalie Loveless argues, "To state the obvious, how one does one's pedagogy in a field impacts what *can* and *is* done in that field."[2]

We may be in a moment where the regimes of modern knowledge are breaking open; they are inadequate to the moment and insufficient to address the exigencies of our time. This book dwells in the potential of this fundamental break.

The educational machine

Bettina Love joins with bell hooks in envisioning education as a practice of freedom. Love's work is about radical collective freedom-building that counters pervasive injustice, and the role our education systems and our knowledge practices have in this pursuit. There is potential for educational spaces to work in solidarity with communities of color intersecting with queer communities, disability communities, and all who are working toward justice. Love calls us toward an abolitionist education characterized by the centrality of the "imagination, creativity, refusal, (re)membering, visionary thinking, healing, rebellious spirit, boldness, determination, and subversiveness of abolitionists to eradicate injustice in and outside of schools."[3] This vision should guide us in the environmental humanities, as our work must be vigorously pushing toward justice.

Yet our situatedness within educational institutions makes our justice work fraught. We are embedded in an institutional context that continues to render its colonialist thinking and colonialist enforcement. These institutions have been meant to instill the idea, in Love's words, that people of color "have had no impact on history or the progress of mankind," which is a foundational idea of white supremacy.[4] This idea plays out both in content and form in educational systems—both in what is taught and how it is taught. This is a point bell hooks makes across her oeuvre: She writes of how "imperialist capitalist white-supremacist patriarchal politics has shaped learning communities, affecting both the way knowledge has been presented to students and the nature of that information." This is a manifestation of what hooks calls dominator culture: the perpetuation of colonial regimes that insist upon imperialist capitalist white supremacist cisheteropatriarchy.[5]

From Patrick Wolfe, we know that colonialism is a continued, perpetuated structure and not a single event.[6] Racism is concurrent with colonialism, or in Wolfe's words, "race is colonialism speaking." The education system has been from the start a part of the colonialist structure and its reproduction. Nominally, our ideals are to educate and to transform, but conventional educational practices work to subordinate and control. As Foucault made clear, disciplines *discipline* and academe is embroiled in many larger systems of control and stratification through debt, policy, and evaluation shrouded under the pernicious myth of a meritocracy and the assumption of "a model of 'uplift.'" Daniel Heath Justice describes this model as the idea that those who are not beneficiaries of the systems we live within "must be lifted up to the social status of the privileged but ostensibly sympathetic activist-observer." He continues,

> In such a model, there's no reflection on whether or not the observer's status quo standards hold any appeal for anyone else and no thought that these standards and values might be considered dangerous or even corrosive to those being "helped."

Carrying this legacy of patronizing arrogance and violence, the classroom remains a place without integrity, as bell hooks has explained. By integrity, hooks means the "integration of ideals, conviction, standards, beliefs—and behavior . . . when ideals and practice match," a definition she borrows from Nathaniel Branden.[7]

The colonial legacies embedded in the university are everywhere apparent. My campus' mission-style architecture is only one manifestation; we can see these legacies in the structure of the institution itself: with its enforced "divisions" of knowledge. The academy is a manifestation of an epistemology that sees value in division and hierarchy,[8] antagonistic to epistemologies that center interconnectedness. This epistemology structures systems of recognition in the university: why Aztec dance is more rarely recognized as a knowledge practice than computer science. Under academe's dominant epistemology, it is a supposed given that nanotechnology should be better resourced and prioritized than the study and practice of basketmaking; this is in willful ignorance of, and derision toward, the rich and varied Indigenous traditions of weaving and the encoding of history and the transfer of knowledge in collective practices of craft that are tied to the land. Academe is inscribed in the colonial manufacture of logics that shape the demand to produce (e.g., research, dollars, or any other unit of productivity) within the academy.

Care for our planet requires that these colonial-capitalist logics be dismantled. We may acknowledge this fact and may even teach about the limits of these logics in our classes, but are we, as a field, striving to dismantle the structures these logics have produced? This question is about dismantling racism and countering white supremacy on and off our campuses. Racism is insidious. The performative valuing of diversity, multiculturalism, and antiracism within university contexts[9] is no cover for the relentless reproduction of racist structures and inequities through many means, including epistemic discrimination. As Rauna Kuokkanen notes, "In the contemporary university, it is no longer non-Western people, but rather their systems of knowledge and their perceptions of the world, that are labelled inferior."[10] It remains sanctioned to speak of the inferiority of cultural knowledges and practices, when other forms of explicit discrimination may be less sanctioned, although nonetheless persistent in practice. Discrimination via the hierarchizing of epistemologies accompanies the racism that is carried out at a structural and interpersonal level. White supremacy continues to be reinforced in subtle and unsubtle ways: in who gets hired, who gets assigned which forms of work, whose work gets elevated by awards committees, which knowledges count as legitimate.

In the environmental humanities, every class we teach is about interlocking oppressions and the effects of intellectual domination, even if we fail to address matters in these terms. As Kyle Powys Whyte notes, following Daniel R. Wildcat,

> climate injustice is a recent episode of a cyclical history of colonialism inflicting anthropogenic (human-caused) environmental change on Indigenous peoples. . . . Climate injustice, for Indigenous peoples, is less about the spectre of a new future and more like the experience of déjà vu.

Climate injustices "began well before the last 250 years of industrial development."[11]

A disciplinary understanding of the environment may enforce hierarchy, as Sarah Jaquette Ray makes clear in *The Ecological Other*. Ray identifies how white environmentalism's framing of "purity and pollution" serves to fortify white supremacy and ableism; environmentalism enforces a social order by making certain bodies virtuous and other bodies expendable.[12]

A parallel to what Ray describes occurs in terms of epistemology: a white tradition is centered as virtuous and relevant to the environmentalist project (e.g., the Thoreau, Leopold, Carson canon) while other cosmologies and epistemologies exist for the environmentalist project only as passive resources to be extracted and assimilated into an existing intellectual repertoire.

White environmentalism's insistence on the separation of nature from culture, environmental concerns from human concerns, continues. As Robert Bullard, Claire Jean Kim, Curtis Marez, Rob Nixon, David Naguib Pellow, Laura Pulido, Julie Sze, and many others have identified, environmental justice concerns are in every environmental concern; yet, the organization of knowledge in the academy does not make this truth central. Instead, biology, botany, and forestry are taught not within frameworks of social justice, and it is often only through the tireless efforts of an Indigenous studies department that Indigenous ecological knowledges may be honored in these courses. Summarizing this reality in western systems of knowledge production, Linda Tuhiwai Smith identifies a colonial orientation that

> assumes Western ideas about the most fundamental things are the only ideas possible to hold, certainly the only rational ideas, and the only ideas which can make sense of the world, of reality, of social life and of human beings. . . . Some indigenous and minority group researchers would call this approach simply racist. It is research which is imbued with an 'attitude' and a 'spirit' which assumes a certain ownership of the entire world, and which has established systems and forms of governance which embed that attitude in institutional practices.[13]

By declaring that "Western ideas about the most fundamental things are the only ideas possible to hold," other worldviews are discredited, marginalized, or eviscerated in a colonial academic milieu. This is epistemic injustice.

The potential of education

Education has served as a central tool in enforcing assimilation. Built to reinforce the hierarchies of power, Aihwa Ong notes that education is "a social technology—in the Weberian sense of appropriate means to an end—for constituting subjects in particular spaces of calculation."[14] There is no turning from this. To be imbricated in educational spaces is to be part of this matrix of power. How we work within educational spaces to shift these relays should be our central concern, and this orientation affects all aspects of our work.

As educational arenas are part of existing apparatuses of power, they also have the potential to transform them. This book offers a critique of educational spaces, but my critique comes from an investment in the potential of these spaces. As la paperson writes in the opening of *A Third University Is Possible*, "Within the colonizing university also exists a decolonizing education." They continue,

> Regardless of its colonial structure, because school is an assemblage of machines and not a monolithic institution, its machinery is always being subverted toward decolonizing purposes. The bits of machinery that make up a decolonizing university are driven by decolonial desires, with decolonizing dreamers who are subversively part of the machinery and part machine themselves.[15]

In fullest acknowledgment of the continued prevalence of racism of white-supremacist and colonialist thought and action in higher education, this trust in the dream and possibility of decolonizing education is persuasive. It is in bell hooks' vision for "academic settings [which] are one of the few locations in our nation where individuals cross the boundaries of race to learn from one another and join in fellowship together."[16] Education spaces are more than one thing because they represent people coming together and as such can be a site of community power. There is radical potential in gathering with intention, and institutions of education provide these opportunities for gathering. However, every time we fall back on the canonized and entrenched, rather than prioritizing what is emergent, we are missing this potential. We need to prioritize what can come into being with the full engagement of those who come to educational spaces.

Aligning with a liberatory vision for education requires a persistent and deep skepticism of educational conventions. In the environmental humanities, we should follow those who are working to intervene in the moment's most dire realities, and we should know that this work is fundamentally stymied by colonialist educational conventions. The educational practices we've inherited prioritize obedience over world-building, conformity over the emergent, and disciplining practices over freedom and autonomy. Not only does obedience reinforce supremacist power dynamics but it is also oriented toward what is, rather than what can be. We need educational environments that foster the creativity and freedom for transformation. Unsettling educational practices becomes an imperative. This book is a thought exercise in this arena, in humble concert with many educators who are exercising their thinking.

We seek imaginations that resist oppressive regimes and build adaptive, responsive multispecies futures. bell hooks notes the necessity of this response: "We are bombarded daily by a colonizing mentality," she writes, "so we must be constantly engaging new ways of thinking and being."[17] The environmental humanities, if the field makes good on its potential, can be oriented toward the dismantling of racial capitalism and the extractivist orientation that undergirds it, exposing the lie of what Enrique Dussel has named the "developmentalist fallacy."[18] Ecological devastation and environmental injustice will not be remedied within the systems that continually perpetuate these harms.

The Quechua word pachakuti is often translated as "world reversal." The word is associated with the powerful mobilization of Indigenous people of the Andes and the anticipated far-reaching upheaval that is necessary to overturn colonial orders.[19] As Mario Blaser explains, the word combines the Quechua word "pacha" (world, time, and space, or state of being) and "kati" (meaning change, turn, or something that comes back to itself).[20] While I have no entitlement to this word, it guides me in my intention to be in support of something that is far beyond me. It guides me to defer to the knowledges and leadership of those who have built and sustained movements and ways of life and epistemologies that have been in shifting alignment with the heartbeat of this earth since time immemorial.

The environmental humanities, as a historically white-dominated field in a particular academic context, can come to follow the lead of those who, in Priscilla Ybarra's words, "never became environmentalists in the first place"[21] because environmentalist orientations have reinforced and reassured white supremacy. Ybarra finds that the epistemologies of Mexican American and Chicana/o cultures have from the start been inclusive of nature and wary of notions of possession; they are thus incompatible with white environmentalist practices and represent a rejection of the exploitive (command and control) epistemologies that characterize the white settler. Ybarra traces profound environmental regard in Chicanx traditions that exceed white environmentalist thought.

The vibrant intellectual productions taking place across the world exemplify rich and varied epistemologies that counter colonial orders through their very existence. For those dehumanized by dominant orders, existence is resistance—existence that includes the epistemologies that sustain lifeways. In their presence, to continue to teach environmental humanities within dominant frames is to perpetuate the objectification that enables violence against the earth and its peoples.

In order to honor the cosmological and intellectual diversity of the world and to dismantle the imposition of colonial orders, this book engages an idea of pluriversality. The term is a resource for both the redress of epistemic harms and engagement with futurity. Pluriversality is often considered in relation to the World Social Forum slogan: "Another world is possible."[22] Other *worlds* are possible, and other worlds have been and continue to be. Today, there are approximately 370 million Indigenous peoples occupying 20 percent of the earth and representing approximately 5,000 distinct cultures, lifeways, and epistemologies. While we share a physical planet, the nature-culture systems that exist on this globe are unquantifiably varied. In cognizance of this reality, the World Social Forum slogan, which was initially written in the singular (another *world* is possible), has, Arturo Escobar notes,

> become more radically pluralized, none the less by social movement mobilizing against large-scale extractive operations in defense of their territories as variable worlds where life is lived according to principles that differ significantly from those of the global juggernaut unleashed on them.[23]

Our field needs to honor these movements and lifeways.

"[R]ealities are plural and always in the making," Escobar reminds, "and this has profound political consequences" (ix). What does it mean to stay near these plural realities-in-the-making in the environmental humanities? And how does one come to dwell among plurality, without converting that dwelling into ownership, co-optation, appropriation, or hierarchy?

Against imposition

Mindfulness is one resource I suggest here, in part because it is itself pluriversal—with innumerable epistemologies that have come under the rubric of contemplative practice, with respect for the heterogeneity of these cultural orientations. Mindfulness, because it offers suspension—of interpretation, judgment, resolution, ownership—offers an epistemological pause, an aperture that creates space between dominant ideas and what transcends them.

The Center for Contemplative Mind in Society defines contemplative practice as "any activity undertaken regularly with the intention of quieting the mind and cultivating deep concentration, calm, and awareness of the present moment."[24] While often secularized in the academy, contemplative pedagogy draws upon many long-standing legacies of practice. Contemplative pedagogy is understood as deriving from the "wisdom traditions" of the many yoga philosophies (Hatha and Tantric, Jnana, Ashtanga, Bhakti) dating back to the Vedic traditions, of ancient practices of Buddhist meditation that turn from a belief in a separate self, metaphysical reflection from the Sufi tradition, and the contemplation of the Jewish Kabbalah, among other lineages.[25]

Contemplative pedagogy has seen increasing interest in the university, being included in the environmental sciences, environmental humanities, and sustainability studies (Eaton, Hughes, and MacGregor, 2017; Sol and Wals, 2015; Wamsler, 2017; Wamsler, Brossman, et al., 2018; Wamsler and Brink, 2018; Wapner, 2016). Eaton, Hughes, and MacGregor's 2017 collection, building upon the Curriculum for the Bioregion Initiative, outlines specific learning goals for sustainability studies and states how contemplative practice is germane to each. These goals include the invitation for students to

- Deepen personal connections to the world around them;
- Navigate the anger, grief, and despair that can arise when highly problematic issues are studied deeply;
- Build capacity for discernment, to better understand problems and their roots, their scale, their interconnectedness, as well as self-discernment with respect to challenging issues and questions;
- Engage multiple perspectives, especially with contested issues;
- Develop a sense of responsibility for the future; make creative responses—as an antidote to paralysis and cynicisms; sustain themselves for long-term engagement.

These learning goals align with a "whole student" approach to pedagogy,[26] as these goals consider mind, body, and spirit and prioritize active, participative, experiential, and integrative pedagogy. These goals affirm the life and humanity of each student and make space for the big questions in life. Behind each of these worthy goals for the environmental humanities course is not only a need for the support of contemplative practice to be present to these concerns; behind each of these are also epistemological questions. What does it mean to deeply connect with the world, to know the world in a way that is connective? How are problematic issues framed within different epistemologies? How does an understanding of a problem's roots, scale, and interconnectedness change when multiple epistemologies are considered? How can we contextualize multiple perspectives as being part of multiple epistemologies and multiple worlds—diverse lifeways that have been carried forward and that have evolved in responsive ways over centuries? How is the idea of responsibility also pluriversal? As Kuokkanen notes: "concepts such as 'responsibility' may have radically different meanings in different epistemes and systems of thought."[27]

The environmental humanities could be a space to teach an orientation to the cultural complexity of this globe, affirming multiple ways of understanding the radical interconnectedness and radical diversity of the earth. bell hooks calls us toward a "commitment to complex analysis and the letting go of wanting everything to be simple." This emphasis on complexity is already a topos of the environmental humanities: we turn to humanistic study in order to hold the complexity of the world. This commitment to complex analysis can also be aligned with an anti-racist practice: as bell hooks notes, "Segregation simplifies; integration requires that we come to terms with multiple ways of knowing, of interaction."[28] hooks points us to a need for epistemic justice to accompany efforts toward equity and inclusion.

This insistence on complexity and epistemic diversity thwarts a one-world universalism that is born of colonial logics. Complexity means, in Arturo Escobar's words, moving

> beyond the certitudes of modernity and the conventional categories that, it is worth underlining, are the very ones used by the institutions perpetuating

> the crisis; the World Bank, the great corporations, most states, organized religions, and also, to a large extent, the academy.

Escobar continues, reflecting on the categories of knowing that are enforced by these institutions, the categories that "replicate the conception of the world held by the powerful . . . in their reports on 'the ways things are,' *as if the world really were that way*."[29] Neocolonialist manipulations that stratify the globe through systems of debt produce these constructions. They enact a form of global white supremacy and stratification that has a direct throughline to former incarnations of colonialism. Escobar's work is an intervention in the "notion of a 'One-World World'" which "is predicated on the West's ability to arrogate to itself the right to be 'the world' and to relegate all other worlds to its rules, to a state of subordination, or to nonexistence."[30] What would it mean to fundamentally destabilize the hegemonic "world of the powerful," or the "one-world world" that, in Marisol de la Cadena and Mario Blaser's words, "has granted itself the right to assimilate all other worlds,"[31] or what Walter Mignolo terms "the diversity of the only option (the Western way)"?[32]

In embracing a pluriversal orientation as a corrective to one-worldism, this book works from the Zapatistas' radical call from the Fourth Declaration of the Lacandón Jungle: "The nation which we are building is one where all communities and languages fit, where all steps may walk, where all may laugh, where all may live the dawn" and what is perhaps the most well-known line from the Fourth Declaration: "Queremos un mundo donde quepan muchos mundos." In the world we want, many worlds fit.[33]

From the Zapatistas' call, a range of scholars have investigated pluriversality as a conceptual resource. Conversations on pluriversality invite criticality about the ways that "knowledge practices . . . are . . . conditioned to reinstate themselves"[34] through ideologies and institutional practices. Pluriversality becomes a lens for rethinking knowledge-making conventions. Even the momentary choice to use the word "we" may unintentionally universalize, cover over difference, and reproduce hegemony. Thus, pluriversality becomes a "lens for our lenses" or a way of noticing how one comes to see, the assumptions that are built into an assertion or knowledge claim. This meta-level double-lensing (i.e., the lens for a lens) allows one to observe how a frame may be operating. Such double-lensing entails bracketing objectivity, as a way of bracketing one-world thinking. Mignolo describes how the bracketing of truth and objectivity is a gesture that is necessary to pluriversal thinking.[35]

The conception of "Earth as pluriverse"[36] is only in part a metaphor. It is out of respect for the complexity of the cosmos, the interacting ecologies of the earth, and radical interconnectedness across differences that pluriversal understandings come forward. As Escobar writes, "Living in accordance with the idea that there are multiple worlds, partially connected but radically different, entails an entirely different ethic of life, of being-doing-knowing."[37] Worlds—ecosystems, movements, cosmovisions—ramify on this shared earth.

This radical acknowledgment of heterogeneity shapes practices. In Blaser and de la Cadena's words, "heterogeneous worldings coming together as a political ecology of practices, negotiating their difficult being together in heterogeneity."[38] The difficulty of "being together" of shared life, or con-vivir, con-viviality is also emphasized in Mignolo: "Conviviality is not holiday, but a hard and relentless effort toward cosmopolitan localism and pluriversal futures."[39] The question of conviviality is a central question of environment.

The present book is an investigation of what these efforts can look like in the environmental humanities classroom. The book takes as its central exigency the forms of epistemic violence and exclusion that are in direct relation to the material atrocities that continue in our world. To address these violences, I seek to follow three threads of potential: First, the book emerges out of scholarship on pluriversality in conjunction with anticolonial, antiracist, intersectional feminisms of color orientations. Second, this text turns to educational praxis and practitioner-based scholarship on culturally responsive pedagogy, funds of knowledge, community cultural wealth, and cultural humility. As a third thread in the braided reference points that this book relies upon, I find resources in the field of contemplative practice.

I considered titling this book "Against Imposition" taking an epigraph from Humberto R. Maturana from 1985 ("*Biologie der Sozialitat*"):

> When one puts objectivity in parenthesis, all views, all verses in the multiverse are equally valid. Understanding this, you lose the passion for changing the other.

These words begin the first chapter of Mignolo's *The Darker Side of Western Modernity*. Western models of education have been figured as the activity of changing the other; the "subject who knows" (teacher) imparting the knowledge that the other (student) is presumed to need. The platitude *I learn as much from my students as they learn from me* does not change the fundamentally patronizing structures of education. The problem is not that education brings about change; the problem is in the saviorist directionality of change, as an instructor is (or thinks of themself as being) charged with the task of changing the other. The power relationship here is the concern, as it was Freire's concern in his critique of the banking model of education.[40] The power relationship remains even in so-called "decentered" classroom spaces. Structures of education remain assimilationist.

I ultimately decided against this title, "Against Imposition," because some things do need to be imposed—the necessity of breaking open the frameworks of white racial logics, settler colonialist perpetuation. This breaking open is done through the work of thought leaders who write, orate, and organize on Indigenous land and in diasporas. The intellectual-epistemic work of these thought leaders stands to give direction to the environmental humanities.

Pluriversal studies

This book is in conversation with pluriversal studies, a field that follows from the Zapatista movement, the uprising of the Ejército Zapatista de Liberación Nacional (EZLN) in Chiapas, Mexico, in 1994. Mignolo has been using the term pluriversality in his work across these three decades, identifying how the rhetoric of modernity and logics of coloniality are two sides of a ribbon running through multiple centuries of continued history: the rhetoric of modernity being the "constantly named and celebrated (progress, development, growth)" top sheen of the ribbon, and colonialism being its underside, disavowed and "silenced or named as problems to be solved by the former (poverty, misery, inequalities, injustices, corruption, commodification, and dispensability of human life)."[41] Pluriversal studies seek to formulate means of releasing the bind of this thread.

A hallmark of pluriversal studies is a trans-comparative approach, reading across multiple cultural formations and lines of thought while also looking for threads and throughlines of a reliably interconnected planet: Arturo Escobar looks to Muntu cosmology, Buddhism, among other intellectual traditions to enact respect for the "multiplicity of mutually entangled and coconstituting but distinct worlds."[42] To escape capital-centric and colonial hegemonies, pluriversal thinkers expound the belief that the "realization of multiple reals/possibles is the best antidote against globalocentric thinking." This realization "enables us to consider the power of place-based and of local becoming in new forms."[43] Thus, a pluriversal approach is always engaged with ancestral continuity, continuance, and reclamation as it is also engaged with emergence and futurity.

Together pluriversal thinkers explicate how, in Boaventura de Sousa Santos' words, "the adequate recognition of injustice and the possible overcoming of oppression can only be achieved by means of an epistemological break."[44] Breaking from genocidal and epistemicidal colonial mentalities requires a reparative economy and approach to mind, body, spirit, and land. In *Epistemologies of the South*, Santos identifies epistemicidal monocultures (sharing in Vandana Shiva's metaphor-not-metaphor "monocultures of the mind") that produce nonexistence in capitalist modernity: specifically, monocultures of knowledge (and accompanying limiting definitions of "rigor" and expertise); monocultures of linear time (and accompanying assumptions about teleologies and disjunctions from tradition); monocultures of the naturalization of differences (naturalized social classification that perpetuate relations of dominance); monocultures of the logic of the dominant scale (e.g., the universal and the global); and monocultures of the capitalist logic of productivity. As hegemonic forces, these monocultures continue what Santos calls the "sociology of absences" that disappear diverse worldsenses in their totalizing vision. These monocultures are designed to eradicate the complex ecologies that value diversity, complexity, and relationality[45]: ecologies of knowledges, ecologies of temporalities, ecologies of recognition, ecologies of trans-scales, and ecologies of productivities. These ecologies counter the modernist assumption of an ontology of separation, thinking instead through a lens of radical relationality, interdependence, and co-constitutive indeterminacy.

An outgrowth of pluriversal studies is what Mario Blaser has called political ontology. Blaser defines political ontology as "a loosely connected project emerging from the convergence of ideas advanced in various scholarly fields," including Indigenous studies, science and technology studies, posthumanism, new materialism, and political ecology, among others.[46] At once a "political sensibility, a problem space, and a modality of analysis and critique," in Blaser's definition, political ontology is neither a pedagogical project nor a blueprint for a good life[47]; instead, political ontology seeks to realize its commitment to the pluriverse through a variety of means: respect for multiplicity, mobilization of multiple ontologies, and story.

My questions in the present text grapple with what pluriversal thinking can mean for teaching, even as pluriversal orientations are anathema to a didactic method. What does it mean to enact a multiepistemic environmental humanities within a classroom space, given the vectors of power that characterize those spaces? What does it mean for the environmental humanities to make good on the core assertion repeated multiple times across Boaventura de Sousa Santos' work: that cognitive and epistemic justice are necessary to social justice?[48]

An epistemic orientation such as what I've outlined cannot neglect the material: the integrity of the body, redistribution of wealth and power, land rights, anti-extractivist recovery and revitalization, self-determination, and more. The false cleaving of binary thinking has been well-countered. The material and the epistemic are not separate. To neglect the epistemic is to neglect the ways in which cultural imaginaries are constitutive. As Escobar notes, "all forms of politics are ontological in that they all involve an ontological dimension: they have implications for what counts as real, for modes of existence, and for adjudicating ethical or nonethical action."[49] A consideration of the ontological and epistemological, then, is necessary for any intervention in political hegemonies; however, to take heed from Kwakwa̲ka̲'wakw scholar Sarah Hunt: "Surely we must engage with the powerful system, but appealing to the law alone won't stop the violence"[50]—solely intervening in the epistemologies that sustain the violence won't stop it either. Perhaps it is enough to think of the attention toward the epistemological and ontological as insufficient but necessary. The lines between epistemological intervention and material change are multidirectional.

As Mignolo writes in the foreword to the collection *Constructing the Pluriverse*, pluriversal studies are grounded in

> decolonial projects emerging out of the global political society (deracializing and depatriarchizing projects, food sovereignty, reciprocal economic organization and the definancialization of money, decolonization of knowledge and of being, decolonization of religion as a way to liberate spiritual, decolonization of aesthetics as a way to liberate esthesis, etc.).[51]

The reclaiming of knowledge, spirit, and being is tied to decolonial projects. Kuokkanen identifies epistemic justice as "the recognition and acknowledgment of the very existence of a multitude of discourses (or epistemes)." She writes,

> Not until hegemonic institutions and individuals have become aware of different discourses, once the multitude of epistemes is made visible, can we proceed with the project of decolonization, of deconstructing Eurocentric biases, or dismantling the hierarchies of discourses and epistemes.[52]

About the text

The argument in this book is inherently incomplete, and I offer it with vulnerability. My intention is to be transparent, before colleagues and students, in how I consider my own practice as a teacher: the questions I am asking, the steps I've provisionally said "yes" to as I work. I write as someone who doesn't have answers, but who nonetheless steps into the classroom day after day with the responsibility to help create or preserve a space for conversation and thoughtful activity.

I write as a white woman who teaches in higher education, a system that perpetuates racism as it maintains its legacy of white supremacy and its entrenched noblesse oblige orientation. The historically white-centering institution, unless it has done significant transformative work, offers a racist education that reproduces inequities for its students.

To become more accountable for this educational situation that I benefit from and am complicit in, I seek to listen more deeply to the voices of activists, including student activists, who have sought to bring higher education, and the larger system it is part of, into a moment of reckoning. This is the ongoing context of my learning, and I make it visible here. I am not a singular mind projecting a vision for how higher education should be remade; this vision should not come from a singular voice—and it should not be my vision, as I remain embedded in the white, colonial logics that reared me. Uprooting those logics is lifelong work. As Paulette Regan writes in *Unsettling the Settler Within*,

> we will always be engaged in the struggle to unlearn racism in our homes, schools, workplaces, and communities, and . . . we will inevitably make mistakes along the way. What makes the difference in terms of a life-long commitment to anti-racism is the willingness to continuously face our mistakes and take the actions necessary to make amends on personal and political levels.[53]

I write with this intention: to engage in the struggle, to keep a commitment. My name is listed in the byline because I seek to be accountable. I am accountable for what it means to teach as mass death continues, caused by the preexisting condition of racism and the forms of brutality it produces in healthcare, policing, housing, education, and innumerable other spheres.

I don't write with answers, but I write with a sense of answerability. I make decisions every day that affect the lives of BIPOC (Black, Indigenous, and People of Color) students and colleagues, due to my situatedness within the institution. We are linked to one another, one action produces effects for another. We each have a role to play in transformation. What Angela Davis said in a 2014 speech at Southern Illinois University is a call to us all: "You have to act as if it were possible to radically transform the world. And you have to do it all the time." I write with this posture and practice: with the pronounced responsibility to seek radical transformation.

I also write as a contemplative practitioner, with a practice of mindfulness as I write. I wrote this book in all seasons: with sun on my face and blankets wrapped around my body. My intention was to write this book with a practice of attention. The buzz of a hummingbird's wings is what I perceive most specifically at this moment. The flying jewel is hovering just three feet behind me. Now it flits away.

With each moment of presence in writing these pages, questions stayed with me about what it means to do scholarship at this moment. I sit with the privilege of my breath moving in and out of my lungs and the stark questions that shape my embodiment in the world: What does it mean for me to participate in scholarship, with all that I don't know? How can I put myself in spaces where I can learn what I need to learn? How can I do better to show up for the work of movement-building? While I want to preserve the possibility of coming to detach from white racial ways of knowing, I also am conscious of how decolonization and deracialization for someone like myself will never be complete. Given this, what does it mean for me to teach as a white person when white faculties are disproportionately represented and empowered, perpetuating continued harms, inequities, and exclusions? The faculty at my campus, a designated Hispanic-serving institution (HSI), is 72 percent white. I am a beneficiary of—and a participant in—the interlocking systems that have for centuries perpetuated segregation, underrepresentation, misrepresentation, and tremendous harm. What does writing and theorizing about the environment through the humanities do in this context, and what does it mean to do this work from my positionality? Referring to Charles Mills' argument, Harvey Cormier writes of how white people—even those who may in some ways be "renegades" will not be able to "see things entirely from a nonwhite point of view."[54] My positionality and the sensibility from which I write is important to draw attention to, in order to demonstrate its limits and the politics of speaking from the subject-position I inhabit. I bring in my "I" in this passage not to center myself, or center whiteness, but to acknowledge the contingency and obstructions of my "I." This is a core idea in the pages that follow, which grapple with questions regarding positionality, interpretation, recognition, and knowledge-making through the framework of epistemic justice.

I write not within a framework of answer-giving, but from the truth that there is no way of opting out of a response. We are always already in response. We respond in our living—in our choices and our actions, in the words we utter or don't. It is from this realization that I came to mindfulness and contemplative practice. In realizing that my life is a response, I also realize that this life calls for mindful attention. In the posture of living the questions, this writing is not set apart from my life.

If I have a case to make here, it is one that lands on a daily practice of not presuming to know, as it also remains firm in the assurance that we must find a revolutionary end to racism and interlocking forms of oppression. I have no final pronouncement here other than that intention, set with acknowledgment of my complicity in sustaining and perpetuating the dehumanizing tendencies of higher education.

The book is aspirational and is not meant to be a pedagogical "success narrative." I want to be clear that the pedagogical discussion that follows is by no means fully realized in the classrooms I facilitate and indeed the ideas described here I may be able to come only ever asymptotically near. While this is a text about the utter complexities of teaching, I have become skeptical of what has elsewhere been called a "recipe-swapping" approach to pedagogy, as if classroom praxis could be outlined in a series of written ingredients and steps. We cannot make the same nourishing soups in each of our classes. We each come to the teaching space with different positionings, identities, ways of being in the world, and relationships to the community of people who make up a class. This makes each educational moment and gathering its own. In this context, there can be no firm pronouncements for the "how" of pedagogy—we can offer only the ability to attend to our practice with a sharper focus, a honed set of questions, and a steady commitment to transformative practice. The questions we ask about the "how" of things are very important: to help us envision and embody a practice. Teaching practices shift and change with every encounter we have with people who stand to teach each other so much. Each classroom space is emergent.

The choices I've made in writing this book come from a consciousness of the significant risks of my position and practice: specifically, as Zoe Todd teaches, there is the risk of flattening, distorting, and erasing forms of thinking that I cannot fully know. My intention is never to render scholars as, in Todd's words, "disembodied representatives of an amorphous Indigeneity that serves European intellectual or political purposes."[55] This colonial orientation toward a generalized Indigeneity elides "the embodied, legal-governance and spiritual aspects of Indigenous thinking"[56] and the unique lifeways and cosmo-epistemologies that may be in coalition with each other, but should not be homogenized.

Vanessa Watts, one of Todd's interlocutors, echoes the point, identifying how Indigenous thought is co-opted and used

> as a gateway for non-Indigenous thinkers to re-imagine their world. In this, our [Indigenous thinkers'] stories are often distilled to simply that—words, principles, morals to imagine the world and imagine ourselves as in the world. In reading stories this way, non-indigenous peoples also keep control over what agency is and how it is dispersed.

This tendency is present, Watts notes, in fields thought by some to be progressive and aware of social justice concerns: critical science studies, ecofeminism. While certain lines of thought in these fields "may serve to change the imperialistic tendencies in Euro-Western knowledge production," Watts writes, "Indigenous histories are still regarded as story and process—an abstracted tool of the West."[57]

With these risks in full view and in humility, I quote often throughout the text, knowing that the most I can contribute to the conversation is to show up, listen, and help others like myself listen too. I quote more than I summarize, cognizant of the risk covering over or obfuscating meaning. Each citation is an attempt to amplify and direct the reader to the much more there is to be read from the scholars listed in this book's bibliography.

I acknowledge the insidious ways in which citational conventions perpetuate racist, ableist, and heteropatriarchal stratification in academic scholarship[58]; white academics and content producers have stolen the intellectual heritage of BIPOC peoples, and reparations have not begun. The examples are many: the supremacist orientation that appropriates and erases Indigenous contribution to the scientific foundations of western research being just one.[59] There is also the matter of reputation and amplification: Donna Haraway's concept of situated knowledge follows from the words of Buchi Emecheta who is much less frequently cited.[60] Silvia Rivera Cusicanqui has noted how Walter Mignolo has obscured and neglected her own and Indigenous people's work.[61] Rosiek, Snyder, and Pratt usefully identify the ways in which white-dominated fields of philosophy of science, new materialism, feminist materialism, posthumanism, post-qualitative research, feminist technoscience studies, and agential realism ignore agent ontologies found in Indigenous studies due to racism and settler-colonial bias. This instance of "talking past" plays out a long-observed reality, identified in Linda Tuhiwai Smith's *Decolonizing Methodologies:* that Indigenous thinkers are excluded from the conversation of scholars, even as issues of Indigeneity may be the so-called object of study. Because discussions of the scientific foundations of western research rarely include mention of the Indigenous contribution to these foundations, Tuhiwai Smith notes, these discussions perpetuate the implicit premise that it "is simply impossible, ridiculous even, to suggest that the object of research can contribute anything."[62] Indigenous peoples become the object of study—the *known*, excluded from the position of the *knower*.

Regrettably, there are no doubt remaining instances in this book of "talking past," in part due to my ignorance of intellectual spheres beyond myopic peer-reviewed journals, including forums of decision-making and coalition-building. I share this example of my complicity here not as a confessional, but as one more example of how rectification is necessary.

I want to also be accountable for the ways in which a book like this risks contributing to the preponderance of "GREAT WHITE HOPES" scholarship that Zoe Todd aptly names as a dominating force in academic spaces. White authors' expressions of our hopes for the fields of study, institutional structures, and pedagogic spaces where we find belonging can be understood as what Eve Tuck and

K. Wayne Yang call settler "moves to innocence."[63] *If I hope hard enough and well enough, I have done my part*, these examples of white scholarship would seem to say, denying complicity and acknowledgment of how white thinkers benefit from the inequitable systems in place. The book before you seeks no false innocence.

The chapters that follow

This book is an extended literature review, drawing from antiracist scholarship, contemplative practice, pluriversal studies, Indigenous thought, Black studies, feminist of color theory, and the environmental humanities. It follows texts like LeiLani Nishime and Kim D. Hester Williams' *Racial Ecologies* in understanding ethnic studies and Indigenous studies as providing the lenses needed to address the most central concerns in the environmental humanities. David Pellow and Julie Sze[64] have likewise examined the interrelationships of environmental studies and ethnic studies, noting tensions as well as intersections. Speaking of the relationships between ecology and social difference, Pellow has put it clearly: "So the point is not to only show how these seemingly disparate issues are 'linked,' but to demonstrate that they cannot be *un*-linked."[65] I argue that the environmental humanities is bereft without the understandings and approaches that are in continual emergence in ethnic studies and Indigenous studies.[66] At the same time, a simple "importation" of these fields' ways of knowing into the environmental humanities is problematic, given the ways that whiteness has configured the environmental humanities. Ybarra in *Writing the Goodlife* chronicles the white-dominated field of ecocriticism noting how Latinx literatures and epistemologies, texts, and knowledges were largely unacknowledged—or when acknowledged were regarded in essentialized ways—in the field for decades, even as these knowledges long predate the emergence of the environmental humanities.

There are unquantifiable lines of thought and cultural traditions that speak to the central exigencies of the environmental humanities. My purpose is to explicate this reality and cue the field to come to terms with its complicity, as it has forwarded secular white supremacist orientations. To make environmental justice central to the field of the environmental humanities necessitates making epistemic justice central to the field.

The synthesis I offer in the pages that follow is not meant to collapse varied orientations into one. Instead, I see this book as an act of weaving different lines of thought together, so that the innumerable threads that exist beyond these pages might be acknowledged and honored. I use the metaphor of weaving in order to preserve the difference in each fiber being woven in—this is not a melding, but is instead a temporary form. The book is meant to point to what is beyond its pages.

This book is for anyone who teaches about the environment, and it may also be relevant to contemplative pedagogy scholars and practitioners beyond that disciplinary locus. The book operationalizes a definition of the environmental humanities that is broad and inclusive, as it also remains grounded in anti-oppressive pedagogies and a prioritization of humanistic inquiry and literacies. I argue for a pursuit of not only interdisciplinary but also multi*epistemic* approaches, underscoring that, in LeiLani Nishime and Kim D. Hester Williams' words, "Environmental exploitation is ideological, and our ability to address environmental degradation depends on our ability to name and resist those ideologies."[67]

Chapter 1 argues for a pluriversal approach as an epistemic demeanor, related to what has been known as cultural humility. The chapter distinguishes pluriversality from normative pluralism, aiming to decenter white-settler thinking and dismantle imperialist capitalist white-supremacist patriarchal politics. The chapter suggests a posture of "not knowing" that is not a passive stance but is active in guarding against any sense of mastery or consuming another. Cultural humility, as a framework, provides the means of tracing a frame as a frame—in other words, cultural and multiepistemic humility enables us to see that our understandings are never neutral, universal, or absolute. Moreover, cultural humility resists "mastery" as a value or aim; it is instead practice-based, in-the-moment, and processual.

Chapter 2 delineates the difference between pluriversality and interdisciplinarity, unpacking the intellectual discriminations of the interdisciplinary academy. This chapter names the suppression of knowledges and, thereby, their knowers, in the white supremacist culture of higher education. The goal is to expose the epistemological limitations of western environmental thought as these limitations criss-cross the disciplines. Interdisciplinarity is therefore not a sufficient response to intellectual discrimination. To what extent do disciplinary practices interrupt and intervene in dominant ideation? And what are the potential political, social, ecological, etc. consequences of our disciplinary stories?

Chapter 3 suggests mindfulness approaches as a way of cultivating attention to what the foregoing discussion means for classroom practice. Following Rhonda Magee, Ruth King, and Resmaa Menakem who have suggested that contemplative practices are germane to anti-oppression work, this chapter asks: What resources are we offering students for dismantling systems of oppression and for navigating the worlds they know in ways that honor cultural practices and prompt further seeking of cultural traditions from a place of self-determination? To be clear, simply adopting mindfulness practices in the classroom is not enough to dislodge and dismantle white supremacy. Attention to the cultural contingency of mindfulness practices as they are embedded in multiple understandings of self-in-world is essential.

Chapter 4 investigates what reading and writing in the classroom might look like when epistemic justice becomes a guiding orientation for our work. The environmental humanities provides an occasion for investigating the narratives that shape our lives and our approaches to what is most urgent. Our literacy practices in the environmental humanities contribute to the cultural imagination, as they also enable a critical investigation of the cultural imaginations we inhabit. What does it mean to open spaces for "thinking otherwise"? What does it mean to interrogate ideological closures and identify how an author may mobilize dominant narratives and assumptions, without insisting on the forms of critique that tear down a text in critical evaluation? Mindfulness practice may offer the environmental humanities internal and collective capaciousness to create room for addressing epistemic injustice and experiences of environmental injustice and for prompting regenerative imagination to meet uncertain futures.

Our literacy practices provide an occasion for multiepistemic humility as we should guard against an assumption of "cultural continuity" between writer and reader. We should not assume that what is true for me is true for another. As readers, we should not assume that we have a full understanding of the texts we encounter. We should read with the cultural humility that resides in difference and seeks to learn what is offered without the demand for knowability—to follow where the text leads, without the demand that the text is incorporable into one's own sense of self or ways of knowing the world, while also remaining open to how the text may prompt an unraveling of dominant assumptions.

The interdisciplinary field of environmental humanities represents aspirations for the potential of humanistic teaching and research to address our world's most urgent intersectional problems: injustices that affect multiple populations, species, and ecologies. Mindfulness practices reveal to us the ways we have of knowing the world—understanding these ways to be contingent, and acknowledging that our existence exceeds those scripts. It is here we find the possibility to know the world in ways that are not predetermined or programmatic and to meet our fellow bodymind-knowers of this planet, across species, with greater respect.

Notes

1 Bettina L. Love, "Teachers, We Cannot Go Back to the Way Things Were," *Ed Week*, 29 April 2020, www.edweek.org/leadership/opinion-teachers-we-cannot-go-back-to-the-way-things-were/2020/04.

2 Loveless continues the point: "As feminist, antiracist, and decolonizing theorists have long taught us, pedagogical ideologies—regimes of truth—configure the parameters of legitimate research questions as well as what counts as rigorous or excellence for both student and teacher. And, in turn, the ways in which internal and national granting boards understand the stakes and parameters of a field." Natalie Loveless, *How to Make Art at the End of the World: A Manifesto for Research-Creation* (Duke UP, 2019) 13.

3 Bettina L. Love, *We Want to Do More than Survive: Abolitionist Teaching and the Pursuit of Educational Freedom* (Beacon, 2019) 2.

4 Love, *We Want to Do More*, 14. "Mankind" here is, I believe, used deliberately to underscore the patriarchy in the white supremacist thinking that Love identifies.

5 bell hooks, *Teaching Critical Thinking: Practical Wisdom* (Routledge, 2009) 29.

As bell hooks notes, "imperialist capitalist white-supremacist patriarchal politics has shaped learning communities, affecting both the way knowledge has been presented to students and the nature of that information" (29). We could add to this list of intersectional oppressions speciesism. David Naguib Pellow notes, "human suffering and social inequality also are sites where . . . pain is intimately linked to the harm visited upon fragile ecosystems and other animals. . . . Environmental justice struggles reveal how power flows through the multi-species relationships that make up life on Earth, often resulting in violence and marginalization for the many and environmental privileges for the few." He continues, "populations are marked for erasure and early death. Such ideological and institutional othering is linked to the more-than-human world as well." David N. Pellow, *What Is Critical Environmental Justice?* (Polity, 2018) 2–3, 17.

6 Patrick Wolfe, "Settler Colonialism and the Elimination of the Native," *Journal of Genocide Research* 8.4 (2004): 387–409.

7 hooks 32. Daniel Heath Justice, Review of *Real Indians: Identity and the Survival of Native America* by Eva Marie Garroutte," (*Wicazo Sa Review* 20.1 (2005): 144. Quoted in Alannah Young Leon, and Denise Nadeau, "Embodying Indigenous Resurgence: 'All Our Relations' Pedagogy," *Sharing Breath: Embodied Learning and Decolonization*, edited by Sheila Batacharya and Yuk-Lin Renita Wong (Athabasca UP, 2018) 74.

8 Thanks to John Johnson for this point.

9 See Sara Ahmed, *On Being Included: Racism and Diversity in Institutional Life* (Duke UP, 2012).

10 Rauna Kuokkanen, *Reshaping the University: Responsibility, Indigenous Epistemes, and the Logic of the Gift* (U of British Columbia P, 2007) 16.

11 Kyle Powys Whyte, "Is It Colonial Déjà Vu? Indigenous Peoples and Climate Injustice," *Humanities for the Environment: Integrating Knowledges, Forging New Constellations of Practice*, edited by Joni Adamson and Michael Davis (Earthscan Publications, 2017) 88, 102. Daniel Wildcat, *Red Alert! Saving the Planet with Indigenosu Knowledge* (Fulcrum, 2009).

12 Sarah Jaquette Ray, *The Ecological Other: Environmental Exclusion in American Culture* (U of Arizona P, 2013) 1.

13 Linda Tuhiwai Smith, *Decolonizing Methodologies: Research and Indigenous Peoples*, 2nd ed. (Zed, 2012) 58.

14 Aihwa Ong, *Neoliberalism as Exception: Mutations in Citizenship and Sovereignty* (Duke UP, 2006) 139.

15 la paperson, *A Third University Is Possible* (U of Minnesota P, 2017). Web.

16 bell hooks, *Teaching Community: A Pedagogy of Hope* (Routledge, 2003) 71. See also la paperson, *A Third University Is Possible* (U of Minnesota P, 2007).

17 hooks. *Teaching Critical Thinking*, 26.

18 Macarena Gómez-Barris, *The Extractive Zone: Social Ecologies and Decolonial Perspectives* (Duke UP, 2017) xvii, xviii. Enrique Dussel, *The Invention of the Americas: Eclipse of "the Other" and the Myth of Modernity* (Continuum, 1995).

19 See Raquel Gutiérrez Aguilar, *Rhythms of the Pachakuti: Indigenous Uprising and State Power in Bolivia* (Duke UP, 2014) which is an account of a "dynamic wave of social potential that affected public life in plural, polyphonic ways. This opened a space-time of Pachakuti. In other words, it produced a social context defined by disrupting what until then has been accepted as a normal part of every day life" (xix). This disruption includes a "rejection of the following basic assumptions: the separation of production and reproduction from everyday life and from the material conditions for social development and management; the delegation of social sovereignty to 'governing' representatives as the foundation for all political activity; and the restriction of the individual and collective creation of values for social well-being to the oppressive constraints of material value and capital" (xxii). It also includes a rejection of the "symbolic structures that recognize certain forms of domination as acceptable; in other words, the relationship of command-obedience entrenched in great hierarchical divisions between genders and ethnicities" (xxiii).

20 Mario Blaser, "Ontological Conflicts and the Stories of Peoples in Spite of Europe," *Current Anthropology* 54.5 (2013): 557.

21 Priscilla Solis Ybarra, *Writing the Goodlife: Mexican American Literature and the Environment* (U of Arizona P, 2016) 7.

22 See Boaventura De Sousa Santos, *The Rise of the Global Left: The World Social Forum and Beyond* (Zed, 2006).

23 Arturo Escobar, *Pluriversal Politics: The Real and the Possible* (Duke UP, 2020) ix.

24 Dan Edwards, et al. *The Activist's Ally: Contemplative Tools for Social Change* (Center for Contemplative Mind in Society, 2007).

25 For more on the legacies from which contemplative practice draws, see Tobin Hart, "Opening the Contemplative Mind in the Classroom," *Journal of Transformative Education* 2 (2004): 29.

26 David Louis Schoem, Christine Modey, and Edward P. St. John. *Teaching the Whole Student: Engaged Learning with Heart, Mind, and Spirit* (Stylus, 2017).

27 Kuokkanen, *Reshaping the University*, 12. Quoting Rodolphe Gashé *Inventions of Difference,* Kuokkanen further explicates the different epistemologies related to the term "responsibility": "A normative definition in the West views responsibility 'as a mechanical application of a framework of rules that simultaneously relieves the subject of the onus of a decision and, hence, all liability. Yet responsibility implies a measured response that can be carried out 'only if the decision is truly a decision, not a mechanical reaction to, or an effect of, a determinate cause. Western liberal notions of responsibility are often constructed as a social Darwinist 'burden of the fittest,' with the benevolent imperialist self cast as 'helping' those less 'fortunate.' In this discourse, responsibility soon becomes nothing more than a duty, as is apparent, for example, in the UN Declaration of Human Responsibilities, which seeks to establish philosophical foundations for a global ethic" (41).
28 hooks, *Teaching Community*, 78.
29 Escobar, *Pluriversal Politics*, 6.
30 Escobar, *Pluriversal Politics*, 14.
31 Marisol de la Cadena and Mario Blaser, *A World of Many Worlds* (Duke UP, 2018) 3.
32 Walter D. Mignolo, *The Darker Side of Western Modernity: Global Futures, Decolonial Options* (Duke UP, 2011) 298.
33 Ejército Zapatista de Liberación Nacional, "Fourth Declaration of the Lacandón Jungle," *The Zapatista Reader*, edited by Tom Hayden (Thunder's Mouth / Nation Books, 2002) 250.
34 de la Cadena and Blaser, *A World of Many Worlds*, 6.
35 Mignolo, *Darker Side*, 52.
36 This phrase is Escobar's in *Pluriversal Politics*, xvii.
37 Escobar, *Pluriversal Politics*, 27.
38 de la Cadena and Blaser, *A World of Many Worlds*, 4.
39 Mignolo, *Darker Side*, 176.
40 Paolo Freire, *Pedagogy of the Oppressed*, 30th Anniversary edition (Continuum, 2000).
41 Mignolo, *Darker Side*, xviii.
42 Escobar, *Pluriversal Politics*, 75.
43 Escobar, *Pluriversal Politics*, xx.
44 Boaventura de Sousa Santos, *Epistemologies of the South: Justice Against Epistemicide*, 2nd edition (Routledge, 2016) viii–ix.
45 Santos, *Epistemologies of the South*, 175.
46 Blaser, "Ontological Conflicts," 548.
47 Blaser, "Ontological Conflicts," 552 and 559.
48 Santos, *Epistemologies of the South*, viii.
49 Escobar, *Pluriversal Politics*, xiv.
50 Quoted in Coulthard, *Red Skin, White Masks*, 102.
51 Mignolo, Foreword to Bernd Reiter, *Constructing the Pluriverse: The Geopolitics of Knowledge* (Duke UP, 2018) xiii.
52 Kuokkanen, *Reshaping the University*, 77.
53 Paulette Regan, *Unsettling the Settler Within: Indian Residential Schools, Truth Telling, and Reconciliation in Canada* (U of British Columbia P, 2010), 23.
54 Harvey Cormier, "Ever Not Quite," *Race and Epistemologies of Ignorance*, edited by Shannon Sullivan and Nancy Tuana (SUNY P, 2007) 63.
55 Zoe Todd, "An Indigenous Feminist's Take on the Ontological Turn: 'Ontology' Is Just Another Word for Colonialism," *Journal of Historical Sociology* 29.1 (March 2016): 7.
56 Todd, "An Indigenous Feminist's Take," 9.
57 Vanessa Watts, "Indigenous Place-Thought and Agency Amongst Humans and Non-Humans (First Woman and Sky Woman Go on a European World Tour!)," *Decolonization: Indigeneity, Education, and Society* 2.1 (2013): 26, 28.

58 See Sara Ahmed, "White Men" (4 November 2014), https://feministkilljoys.com/2014/11/04/white-men/. See also Katherine McKittrick, *Dear Science and Other Stories* (Duke UP, 2021).
59 Tuhiwai Smith, *Decolonizing Methodologies*, 63.
60 This is noted in Katherine McKittrick's *Dear Science and Other Stories* (Duke UP, 2021).
61 Silvia Rivera Cusicanqui, *Ch'ixinakax Utxiwa: On Decolonising Practices and Discourses* (Polity, 2020).
62 Tuhiwai Smith, *Decolonizing Methodologies*, 63, 64.
63 Tuck and Yang define this term as follows: "Settler moves to innocence are those strategies or positioning that attempt to relieve the settler of feelings of guilt or responsibility without giving up land or power or privilege, without having to change much at all. In fact, settler scholars may gain professional kudos or a boost in their reputations for being so sensitive and self-aware" (Eve Tuck and K. Wayne Yang. "Decolonization Is Not a Metaphor," *Decolonization: Indigeneity, Education, & Society* 1.1 (2012): 10).
64 Here I am referencing the "Building and Fighting at the Crossroads of Ethnic Studies and Environmental Studies panel, presented at the American Studies Association Annual Meeting in Hawaii, November 2019. While I was not in attendance at this panel discussion, David N. Pellow's excellent paper "Environmental Studies and Ethnic Studies: Tensions, Collaborations, Intersections" was shared with me by Sarah Jaquette Ray. In this paper, Pellow references Julie Sze's talk, given at Columbia University, titled "Environmental Justice and Ethnic Studies: The Puzzling Split." In "What Is Critical Environmental Justice?" Pellow links his work with fields such as critical race theory, critical race feminism, ethnic studies, gender and sexuality studies, political ecology, anti-statist/anarchic theory, ecological feminism, queer theory, and disability studies.
65 Pellow, "Environmental Studies and Ethnic Studies," 2.
66 Dian Million makes the important point that these fields should be understood as distinct. She clarifies: "I posit that environmental racism and environmental justice may have important overlapping interests with Indigenous interests, but we cannot go forward without acknowledging their profound difference. There is a shift in relations between working from the perspective of Indigenism to working against environmental racism and for environmental justice. . . . I want to take those differences, differences in epistemologies (ways of knowing) and ontologies (ways of being) into account."
67 LeiLani Nishime and Kim D. Hester Williams. *Racial Ecologies* (U of Washington P, 2018).

Works cited

Ahmed, Sara. *On Being Included: Racism and Diversity in Institutional Life*. Duke UP, 2012.

Blaser, Mario. "Ontological Conflicts and the Stories of Peoples in Spite of Europe." *Current Anthropology* 54.5 (2013): 547–568.

Blaser, Mario, and Marisol de la Cadena. "Introduction: Pluriverse: Proposals for a World of Many Worlds." *A World of Many Worlds*, edited by Marisol de la Cadena and Mario Blaser. Duke UP, 2018.

Bullard, Robert D. *Dumping in Dixie: Race, Class, and Environmental Quality*. 2nd ed. Westview, 1994.

Cormier, Harvey. "Ever Not Quite." *Race and Epistemologies of Ignorance*, edited by Shannon Sullivan and Nancy Tuana. SUNY Press, 2007.

Cusicanqui, Silvia Rivera. *Ch'ixinakax Utxiwa: On Decolonising Practices and Discourses*. Polity, 2020.

Davis, Angela. *Untitled Lecture Delivered at Southern Illinois University*. Carbondale, IL. 13 February 2014. www.youtube.com/watch?v=6s8QCucFADc

Davis, Heather, and Zoe Todd. "On the Importance of a Date, or Decolonizing the Anthropocene." *ACME: An International Journal for Critical Geographers* 16.4 (2017): 761–780.

Dussel, Enrique. "Transmodernity and Interculturality: An Interpretation from the Perspective of Philosophy." *Journal of Peripheral Cultural Production of the Luso-Hispanice World* (Spring 2012): 28–58.

Eaton, Marie, Holly J. Hughes, and Jean MacGregor. *Contemplative Approaches to Sustainability in Higher Education*. Routledge, 2017.

Edwards, Dan, et al. *The Activist's Ally: Contemplative Tools for Social Change*. Center for Contemplative Mind in Society, 2007.

Ejército Zapatista de Liberación Nacional. "Fourth Declaration of the Lacandón Jungle." *The Zapatista Reader*, edited by Tom Hayden. Thunder's Mouth/Nation Books, 2002.

Escobar, Arturo. *Pluriversal Politics: The Real and the Possible*. Duke UP, 2020.

Essed, Philomena. *Understanding Everyday Racism: An Interdisciplinary Theory*. Sage Publishing, 1991.

Fanon, Frantz. "Racism and Culture." *Toward the African Revolution*. Grove Press, 1967, pp. 31–44.

Gade, Anna. *Muslim Environmentalism*. Columbia UP, 2019.

Gómez-Barris, Macarena. *The Extractive Zone: Social Ecologies and Decolonial Perspectives*. Duke UP, 2017.

Grove, Jairus. "Response to Jedediah Purdy." *Forum: The New Nature. Boston Review*. 11 January 2016.

Hart, Tobin. "Opening the Contemplative Mind in the Classroom." *Journal of Transformative Education* 2 (2004): 28–46.

hooks, bell. *Teaching Community: A Pedagogy of Hope*. Routledge, 2003.

hooks, bell. *Teaching Critical Thinking: Practical Wisdom*. Routledge, 2009.

Horkheimer, Max. "Traditional and Critical Theory." *Critical Theory: Selected Essays*. Continuum, 1975, pp. 188–244.

Hunt, Sarah. "More than a Poster Campaign: Redefining Colonial Violence." *Decolonization: Indigeneity, Education and Society* blog. 14 February 2013.

Justice, Daniel Heath. Review of *Real Indians: Identity and the Survival of Native America* by Eva Marie Garroutte." *Wicazo Sa Review* 20.1 (2005): 201–203.

Kim, Claire Jean. *Dangerous Crossings: Race, Species, and Nature in a Multicultural Age*. Cambridge UP, 2015. Print.

Kuokkanen, Rauna. *Reshaping the University: Responsibility, Indigenous Epistemes, and the Logic of the Gift*. U of British Columbia P, 2007.

la paperson. *A Third University Is Possible*. U of Minnesota P, 1997.

Leon, Alannah Young, and Denise Nadeau. "Embodying Indigenous Resurgence: 'All Our Relations' Pedagogy." *Sharing Breath: Embodied Learning and Decolonization*, edited by Sheila Batacharya and Yuk-Lin Renita Wong. Athabasca UP, 2018.

Love, Bettina L. *We Want to Do More than Survive: Abolitionist Teaching and the Pursuit of Educational Freedom*. Beacon, 2019.

Love, Bettina L. "Teachers, We Cannot Go Back to the Way Things Were." *EdWeek*. 29 April 2020. www.edweek.org/leadership/opinion-teachers-we-cannot-go-back-to-the-way-things-were/2020/04

Loveless, Natalie. *How to Make Art at the End of the World: A Manifesto for Research-Creation*. Duke UP, 2019.

Magee, Rhonda. *The Inner Work of Racial Justice: Healing Ourselves and Transforming Our Communities Through Mindfulness*. Penguin, 2019.

Malm, Andreas. "The Origin of Fossil Capital: From Water to Steam in the British Cotton Industry." *Historical Materialism* 21.1 (2013): 15–68.

Marez, Curtis. *Farm Worker Futurism: Speculative Technologies of Resistance*. U of Minnesota P, 2016.
Melamed, Jodi. *Represent and Destroy: Rationalizing Violence in the New Racial Capitalism*. U of Minnesota P, 2011.
Mignolo, Walter. *The Darker Side of Western Modernity: Global Futures, Decolonial Options*. Duke UP, 2011.
McKittrick, Katherine. *Dear Science and Other Stories*. Duke UP, 2020.
Million, Dian. *Therapeutic Nations: Healing in an Age of Indigenous Human Rights*. U of Arizona P, 2014.
Mills, Charles W. *The Racial Contract*. Cornell UP, 1997.
Mirzeoff, Nicholas. "It's Not the Anthropocene, It's the White Supremacy Scene, or, The Geological Color Line." *After Extinction*, edited by Richard Grusin. U of Minnesota P, 2016.
Moore, Jason. *Capitalism in the Web of Life: Ecology and the Accumulation of Capital*. Verso, 2015.
Nishime, LeiLani, and Kim D. Hester Williams. *Racial Ecologies*. U of Washington P, 2018.
Nixon, Rob. *Slow Violence and the Environmentalism of the Poor*. Harvard UP, 2011.
Ong, Aihwa. *Neoliberalism as Exception: Mutations in Citizenship and Sovereignty*. Duke UP, 2006.
Pellow, David N. *What Is Critical Environmental Justice*? Polity, 2018.
Powell, Malea. "Stories Take Place: A Performance in One Act." *College Composition and Communication* 64.2 (2012).
Pulido, Laura. *Environmentalism and Economic Justice: Two Chicano Struggles in the Southwest*. U of Arizona P, 1996.
Ray, Sarah Jacquette. *The Ecological Other: Environmental Exclusion in American Culture*. U of Arizona P, 2013.
Razack, Sharene. *Looking White People in the Eye: Gender, Race, and Culture in Courtrooms and Classrooms*. U of Toronto P, 2006.
Razack, Sherene. *Dying from Improvement: Inquests and Inquiries into Indigenous Deaths in Custody*. U of Toronto P, 2015.
Regan, Paulette. *Unsettling the Settler Within: Indian Residential Schools, Truth Telling, and Reconciliation in Canada*. U of British Columbia P, 2010.
Reiter, Bernd. *Constructing the Pluriverse: The Geopolitics of Knowledge*. Duke UP, 2018.
Rilke, Rainer Maria. *Letters to a Young Poet*, translated by M.D. Herter Norton. Norton, 1993.
Roshanravan, Shireen. "Motivating Coalition: Women of Color and Epistemic Disobedience." *Hypatia* 29.1 (Winter 2014): 41–58.
Santos, Boaventura de Sousa. *Epistemologies of the South: Justice Against Epistemicide*. 2nd ed. Routledge, 2016.
Smith, Linda Tuhiwai. *Decolonizing Methodologies*. 2nd ed. Zed Books, 2012.
Subcomandante Insurgente Marcos. *Our Word Is Our Weapon: Selected Writings*, edited by Jauna Ponce de León. Seven Stories Press, 2001.
Sze, Julie. *Environmental Justice in a Moment of Danger*. U of California P, 2020.
Todd, Zoe. "An Indigenous Feminist's Take on the Ontological Turn: 'Ontology' Is Just Another Word for Colonialism." *Journal of Historical Sociology* 29.1 (March 2016): 4–22.
Tuck, Eve, and K. Wayne Yang. "Decolonization Is Not a Metaphor." *Decolonization: Indigeneity, Education, & Society* 1.1 (2012): 1–40.

Watts, Vanessa. "Indigenous Place-Thought and Agency Amongst Humans and Non-Humans (First Woman and Sky Woman Go on a European World Tour!)." *Decolonization: Indigeneity, Education, and Society* 2.1 (2013): 20–34.

Whyte, Kyle Powys. "Is It Colonial Déjà Vu? Indigenous Peoples and Climate Injustice." *Humanities for the Environment: Integrating Knowledges, Forging New Constellations of Practice*, edited by Joni Adamson and Michael Davis. Earthscan Publications, 2017.

Wildcat, Daniel. *Red Alert! Saving the Planet with Indigenous Knowledge*. Fulcrum, 2009.

Ybarra, Priscilla Solis. *Writing the Goodlife: Mexican American Literature and the Environment*. U of Arizona P, 2016.

1

PLURIVERSALITY AND MULTIEPISTEMIC HUMILITY

As Boaventura de Sousa Santos is well known for saying across his many books: "there can be no global social justice without global cognitive justice." Cognitive and epistemic justice move beyond common models of distributive and participatory justice, seeking justice for the life knowledges and lifeways of being of historically marginalized peoples.

Epistemic injustice may be most readily thought of as the means by which someone is made to be systemically uncredible. The injustice is that which enables the aprioristic rejection of testimony or the dismantling of cultural systems.[1] This rejection can be a form of denial, dismissal, or unhearing that is at once interpersonal and structural. Epistemic injustice includes the ways in which people may be disbelieved when speaking of racialized, gendered, or ableist harms, although epistemic injustice extends beyond this all too common tendency. José Medina defines epistemic injustice by its manifestations: "such as unequal access to and participation in knowledged practices, vitiated testimonial dynamics, phenomena of hermeneutical marginalization," with testimony here being "any kind of telling in and through which the expression and transmission of knowledge becomes possible."[2] Spivak's question "Can the subaltern speak?" is the question: what conditions enable the unhearing?[3]

Many knowledges and ways of being remain unheard in environmental humanities classrooms. Consider, for example, the Indigenous Principles of Just Transition (2018) presented by the Indigenous Environmental Network,[4] an alliance of Indigenous peoples who advocate for the adherence to Indigenous knowledge and natural law with a focus on Turtle Island. The Principles of Just Transition document foregrounds Indigenous prophecy. The earth is out of balance, a result of occupation and colonization and the necessity of living in accordance with Indigenous peoples' respective Original Instructions which were given to Indigenous peoples at the time of creation. The principles of just transition outlined

DOI: 10.4324/9781003037439-2

in this document affirm Indigenous responsibility, duty, and respect to sacred creation principles and natural laws of Mother Earth and Father Sky. A just transition framework is bereft without recognition of the epistemologies represented in this document, but how often is climate change taught in environmental humanities courses without accountability to these principles and the communities who authored them?

The relegation of ways of knowing is a manifestation of white supremacist thinking that is taken for granted. As Charles Mills notes in *The Racial Contract*, white-authored textbooks and courses have taken white racial privilege "so much for granted that they do not even see it as *political*, as a form of domination." He continues, "the system of domination by which white people have historically ruled over and, in certain important ways, continue to rule over nonwhite people—is not seen as a political system at all. It is just taken for granted; it is the background against which other systems, which we *are* to see as political, are highlighted."[5] This dominating structure goes unnamed, uninterrogated. White supremacy becomes so assumed that it becomes invisible, the hegemonic background of knowing—against which multiple ways of being are already undermined. As Mills notes,

> Native American, African American, and Third and Fourth World political thought, historically focused on issues of conquest, imperialism, colonialism, white settlement, land rights, race and racism, slavery, jim crow, reparations, apartheid, cultural authenticity, national identity, *indigenismo*, Afrocentrism, etc. . . . hardly appear in mainstream political philosophy, but they have been central to the political struggles of the majority of the world's population.[6]

What Mills says of the field of philosophy here also applies to the environmental humanities. When the environmental humanities take up questions of nature, climate, animality, and materialism in ways that divest from, leave unmentioned, or push to the margins the forms of knowing that center the experiences, histories, and knowledges of those who have suffered the effects of world systems, the field reaffirms its oppressive legacy. Epistemic justice requires interrogating the terms by which knowledges are considered legitimate, the terms by which the exigencies are determined. Epistemic justice requires undermining the systems that sabotage Black, Indigenous, Third and Fourth World thought.

Miranda Fricker identifies the terms by which one's capacities as a knower become undermined, focusing on two forms of epistemic injustice: testimonial injustice (injustice in who becomes enabled to speak, who is credible as a knower) and hermeneutical injustice (injustice in how one is understood against the interpretive regimes that preexist them). Notably, Fricker's text has been criticized for enacting that which it names: itself representing a form of epistemic injustice (or what Nora Berenstain describes as "white feminist structural gaslighting")[7] in both Fricker's assumptions and her citation practices: she omits the scholarship of Charles Mills, George Yancy, and additional philosophers of color who have theorized the racialized politics of epistemic justice.[8]

Epistemic injustice is a profound violence that ricochets into other violences: the denial of one's existence, the denial of the grievability of a life, and the denial of one's rights to the resources that enable their lifeways. Vandana Shiva's work has focused on how epistemic injustice is directly material, translated into damaged food systems and the invasion of sovereignties. Using the term "monoculture," which refers to both the industrial farming practice and the insistence on the elevation of a singular dominant culture, Shiva notes, "monocultures first inhabit the mind, and are then transferred to the ground."[9] She, with Boaventura de Sousa Santos and many others, writes of the epistemic injustice committed against many knowledge systems of the world through the monopoly of science and the technologies it sanctions.[10] As Shiva writes in *Monocultures of the Mind*, knowledge systems are made to disappear through the negation of their existence, through the insistence on globalized and totalizing knowledge regimes under neoliberalism,[11] and through the supremacist thinking built into these knowledge regimes. Local knowledges are made to disappear, she writes, "by denying . . . the status of a systematic knowledge, and assigning it the adjectives of 'primitive' and 'unscientific.'" This intellectual colonization is carried out not only by the so-called "green economy" and developmentalism but also through the maintenance of academic systems that produce and validate knowledges. Shiva writes of how the formation of western science is poorly equipped for self-examination and the dismantling of its own discriminations. Western, dominant science makes itself absolute and therefore unquestionable, proscribing and making taboo any consequential reevaluation of its fundamental methods.[12] Western science makes itself sacrosanct, as it denies what other knowledge systems make sacred. This is the maneuvering of epistemic injustice.

Accountability to what Santos, Medina, Mills, and Shiva identify as epistemic injustice must be central to the practice of the environmental humanities. The environmental humanities is imbricated in these systems of injustice and at the same time has the responsibility to continually intervene in them. Despite intentions, the field, along with overlapping fields of sustainability studies, critical science studies and science and technology studies, exist inside of the assumptions of dominant knowledge. There is no "cure-all" to be found in a single discipline, especially when that discipline of thought was formed within the western knowledge system of the academy. Participation in these fields does not equate to a decolonizing approach that validates the vast knowledge diversity of the globe, inseparable from the lands from which these knowledges emerged, nor do these fields exist in a space apart from the racial contract that benefits white people. The racial contract, named and described by Mills, is an epistemological contract "between those categorized as white *over* the nonwhites, who are thus the objects rather than the subjects of the agreement."[13] The contract creates "a cognitive and moral economy psychically required for conquest, colonization, and enslavement" and the active sustaining of continued systems of oppression.[14]

We are called to attend to the terms and theories we mobilize within the environmental humanities. As Métis anthropologist Zoe Todd notes, "terms and theories can act as gentrifiers in their own right."[15] Heather Davis and Zoe Todd illustrate this point, taking as an example the term Anthropocene—that disciplinary crossover term attributed to Paul Crutzen and Eugene Stoermer. Davis and Todd make the argument that the term is itself inscribed within colonial thinking, noting the "failure of the Anthropocene, as a concept to adequately account for power relations."[16] By flattening all of humanity into the prefix "Anthropo," the term obfuscates disparities in who has contributed to—and who now bears the most brutal effects of—climate change. In turn, the term makes it difficult to see that, in Davis and Todd's words, "ecocide logics that now govern our world are not inevitable or 'human nature.'"[17] Citing Swedish scholars Andreas Malm and Alf Hornborg, Todd shows how "the current framing of the Anthropocene blunts the distinctions between the people, nations, and collectives who drive the fossil-fuel economy and those who do not." She continues, "The complex and paradoxical experiences of diverse people as humans-in-the-world, including the ongoing damage of colonial and imperialist agendas, can be lost when the narrative is collapsed to a universalizing species paradigm" (as "anthropo") that emphasizes the human species rather than the systems that have made some the beneficiaries of others' exploitation.[18]

Not only the naming but also the dating of the Anthropocene is problematic. The issue of periodization is much debated in the literature, but Davis and Todd insist on the necessity of acknowledging continued legacies of colonialism in any drawing of a timeline. "[T]he Anthropocene is not a new event," they write, "but is rather the continuation of practices of dispossession and genocide."[19] They continue, "the Anthropocene, and the uneven impacts on the global poor, can be understood not just as an unfortunate coincidence or accident, but rather as a *deliberate extension of colonial logic*."[20] The terms we use should foreground, rather than cover over, this reality.

As Davis and Todd note, several terms have been suggested as alternatives to the Anthropocene: Capitalocene, Eurocene, White Supremacy Scene, Chthulucene.[21] These are useful terms, but their use may still serve to neutralize, or mark as inevitable, the capitalist logics, eurocentrism, and white supremacy that the terms name. The assumption that these realities are just a "given of our era" may weaken a capacity for transformation and for redressing the colonial past.

Anna Gade, in a text on Muslim environmentalisms, further extends the critique of the term as she accuses Anthropocene thinking of "ethical abdication" as it mobilizes "cold scientific periodization on the one hand and emotional resorts of hope and alarm on the other."[22] This ethical abdication, in Gade's view, produces the existential crisis that she describes as a

> struggle with the apprehension of precarity—political, sentimental, existential—that no longer represents environmental problems on a scale of manageability, but rather on overwhelming magnitudes of crisis (late market capitalism, nuclear, and/or climate disaster, survival and well-being of populations, mass species extinction, ocean and earth systems in collapse, and so forth).[23]

This particular crisis in secular environmental thought manifests the inheritance of colonialism. Colonialism sets the course that now produces this crisis; the limits of the logics of "command and control" are being exposed.

The logics of colonialism remain deeply embedded in environmental discourses. Zoe Todd reads these logics in climate change discourse, specifically in constructions of climate as *aer nullius*. Todd introduces the term *aer nullius* as a way of naming the colonial "destroy and replace" orientation that is imposed not only on the land but also on the climate. Developmentalist models of climate that prioritize isolated techno-fixes, carbon capture, and geoengineering deny climate as interwoven into diverse ecological systems inclusive of human beings; these technologized "solutions" are an extension of colonial thinking that sees air as *resource*, not as *relationship*. In this thinking, the atmosphere becomes a frontier belonging to no one and waiting to be saved, with the given being that the savior is a colonial power. The illogic of this orientation—knowing that it is western capital expansion that continues to perpetuate the climate crisis—does not slow its march in the direction of techno-remedies bought and sold as free market commodities.

To intervene, academic discourse requires more capacity to question how knowledges are shaped. If our intention as a field is to become more broadly aware of the radical diversity and interdependency of our world—as we work to intervene in the forces that fail to offer each worldly participant due respect and dignity—then our work is in part to understand how those forces of failure may move through us. What the environmental humanities may be best equipped to do is uproot the devastating, yet contingent, epistemologies that sustain planetary and social harms. The dominant systems are contingent in that they are historical, and, in Shiva's words, they are parochial: she writes,

> the dominant system is also a local system, with its social basis in a particular culture, class and gender. It is not universal in an epistemological sense. It is merely the globalised version of a very local and parochial tradition.[24]

How can we leverage the praxis of humanistic study to demonstrate the contingency of dominant knowledge regimes so that they may be dismantled? How can we orient the humanities toward the radical pluriversality of our globe? This pluriversality is in a reciprocal relationship with biodiversity. As Shiva writes, "Diverse ecosystems give rise to diverse life forms, and to diverse cultures. The co-evolution of cultures, life forms and habitats has conserved the biological diversity on this planet. Cultural diversity and biological diversity go hand in hand."[25] The environmental humanities can ensure that this truth is respected, to join the robust movements and cultures that daily assert this reality against the hegemonic and monopolizing forces that seek to cover the globe.

The analytical capacity of the humanities readies us for an examination of the assumptions built into master narratives. Within the frameworks and practices prioritized in the humanities—close reading, exposition of systems of thought, explication of the politics of representation, and so on—we may be equipped, for example, to identify neoliberal logics and trace how these logics are embedded in a particular discourse. To be clear, the suggestion that humanistic study may offer key tools for calling into question the givens of our world does not deny the deep embeddedness of humanistic thought in systems of dominance or its complicity in epistemic injustice. However, the practices associated with humanistic study (analysis of the situatedness of knowledge, metadiscursive thought, interrogation of a social imaginary, cultural production, etc.) have been utilized by practitioners representing multiple ways of knowing. That said, the particular configuration of the humanities within the academy has been fused with particular colonial "enlightenment" ideas of *humanitas* and who counts as "Man." This history must be accounted for in any mobilization of humanistic study, even as we hold out the potential for the humanities to do counterhegemonic work.

In the humanities, we may consider ourselves ready to critique the systems within which we are embedded, but uprooting white supremacist knowledge regimes is ongoing and thorough work, requiring a persistent set of questions: What "ontological designs"[26] do we hold in our classes and in our scholarship? How do we design our thinking in ways that are radically respectful of diverse ways of being on this globe? What onto-epistemologies do we remain close to and at what risk? And what onto-epistemologies can we better respect, better witness, better learn from, without a demand to consume, own, know? An environmental humanities that respects pluriversality and that dismantles extractive hegemonic systems is possible.

Against normative pluralism

Afro-Caribbean scholar M. Jacqui Alexander defines pedagogy as "the imperative of making the world in which we live intelligible to ourselves and to each other."[27] Here, she is pointing to the reciprocal investments we must make in order to cross into a metaphysics of interdependence, into an epistemology that prioritizes relationality in difference. The type of intelligibility Alexander describes requires a multitude of cultural understandings, a cognizance that remains persistently attentive to the racialization and colonization written into the moment of relation in order to enact a form of resistance. Thus, Alexander's discussion of pedagogy is not a simple act of recognition. It subverts easy pluralism, what Ann DuCille in *Skin Trade* calls "the academic merchandising of different difference."[28] DuCille is critiquing the tokenizing and flattening approaches of official discourses that mobilize trite themes of diversity and equality while keeping gross disparities in place.

Chandra Mohanty defines this normative pluralistic[29] approach as the assumption that "we all occupy separate, different, and equally valuable places." Normative pluralism also has the essentializing orientation of defining experience "not in terms of individual qua individual, but in terms of an individual as representative of a cultural group." This orientation, she writes, only serves to depoliticize and dehistoricize as it reaffirms the status quo "in the name of cooperation and harmony."[30] Mohanty critiques the "additive approach" to inclusion that she finds in higher education: "recruiting 'diverse' people, introducing 'different' curriculum units while engaging in teaching as usual—that is, not shifting the normative culture-vs.-subcultures paradigm."[31] Mohanty names the pedagogical challenge we face by posing the question:

> how do we undermine the notions of multiculturalism as melting pot, or multiculturalism as cultural relativism that so permeate U.S. consumer culture and that are mobilized by the corporate academic as a form of containment, and practice a multiculturalism that is about the decolonization of received knowledges, histories, and identities, a multiculturalism that foregrounds questions of social justice and material interests, which actively combats the hegemony of global capital?[32]

Melting pot multiculturalism and an easy cultural pluralism is compatible with hegemonic systems, as it does nothing to change uneven power relations. That which is centered remains centered in what María Lugones calls "the ideology of the ethnocentric racial state which privileges the dominant culture as the only culture to 'see with.'"[33] This ethnocentrism may invite multiple points of view only to assimilate, gaze upon, neutralize, and ultimately obliterate them.

These "valorizations of melting-pot multiculturalism's sameness," or sameness-in-difference, to use AnaLouise Keating's words in the book *Transformation Now!*,[34] are ultimately performative as they also represent a form of colonial assimilationist expansion—taking up the "other" into cannibalizing knowledge regimes. The presentation of a benign coexistence obfuscates the continued genocides of colonialism.[35]

To seek epistemic justice is to reject assimilationist "melting pot multiculturalism" that tokenistically acknowledges cross-cultural understanding while keeping dominant ideologies firmly in place. Multiepistemic practices must resist normative pluralism that reduces and contains knowledges and identities. Pluralism, operating out of a dominant ideology, arrests worldviews into discrete, essentialist units, locking them in time. The move to "lock in time" is born of the settler-colonial elimination narrative: creating the notion that Indigenous worldsenses are "gone" from the modern world, relegated to history. As Rauna Kuokkanen notes, "Worldviews, ontologies, cosmologies, values, and systems of knowledge are dynamic and constantly evolving, so it is impossible to define an indigenous episteme (or more specifically, a Sami, Cree, Inuit, or Salish episteme)."[36] Epistemes are not collectable as knowledges that can be incorporated into dominant frames. Thus, normative pluralism's additive model is corroding. The reductive act of adding cosmologies to existing western knowledges denies their incommensurability and ignores the way power relations are maintained. Normative pluralism enables a nominal "respect for diversity" to replace the necessary redress of persistent intersectional oppressions. "Respect for" is not enough. "Mere respect tends to create a climate of 'repressive tolerance,'" Kuokkanen notes, "in which indigenous people and their epistemes are allowed to exist in the celebratory spirit of different perspectives or points of view but are not recognized, heard, or understood even superficially and relativistically."[37] To truly recognize these epistemes would fundamentally destabilize the order of the world, as these epistemes confront the frameworks and hierarchies dominant systems are built upon. When this decolonizing form of recognition does not occur, the result is a form of tolerance that amounts to repression, as ways of knowing continue to be marginalized. This is a form of violence shrouded, but not hidden, in liberal discourses of appreciation for diversity and tolerance.

Jodi Melamed[38] identifies how repressive tolerance and normative pluralism have been central to the apparatus of humanistic education. In *Represent and Destroy*, she identifies three successively prominent versions of what she terms "official antiracism"—which is antiracist in name only and which serves to deter radical transformation in its performativity.[39] Focusing on post-war US policy and practice, Melamed diagnoses the formal or official antiracisms that perpetuate the logics of white supremacy under a lexicon that includes terms and tropes such as race reform, racial progress, inclusion, and more.

Melamed outlines three periods associated with three versions of official antiracism: First, the racial liberalism of the mid-1940s to 1960s was a "framework that conceived of racism as prejudice and promised to release liberal freedoms from racial restrictions by extending equal opportunity, possessive individualism, and cultural citizenship."[40] These liberal discourses—while circulating "under the motto of creating change for the good of society"—shored up the status quo and reaffirmed hierarchized relations: "the superior oppressor and inferior oppressed."[41] We can turn to bell hooks to further elucidate this point: in *Teaching Community*, hooks reflects on how "many people supported inclusion only when diverse ways of knowing were taught as subordinate and inferior to the superior ways of knowing informed by Western metaphysical dualism and dominator culture."[42] Projects meant merely to extend access to equal opportunity, possessive individualism, and citizenship—launched under a patina of white "benevolence"—can only reinstate the disparity between the giver and the given, as they also "colonize the consensual real" (Melamed's apt phrase) with the hegemonic orders that are "inevitably reductive and violating"[43] to those whose lives are acknowledged only as a second step, with the rights of some extended to "others" only secondarily.

In Melamed's periodization, the liberal multiculturalism of the 1980s to 1990s brings about the canon wars in the humanities. Liberal multiculturalism's official antiracism emphasized representation and the inclusion of diverse voices while omitting the continued realities of racial violence from the presentation. Under the rubric of "representation, authenticity and gaining voice," liberal multiculturalism insisted on a dematerialized conception of race, ignoring its reinstitution of differential power.[44] Melamed writes of this form of official antiracism:

> As multicultural ideas of integration, representation, and recognition in culture (the last narrowly construed as aesthetics) became the horizon for knowing race and antiracism, they deflected attention from the devastating and

> accelerating consequences of private capital's economic prerogatives for black and brown lives.[45]

Again, in this second configuration in Melamed's timeline, perfunctory gestures of recognition are instrumentalized in the dominant culture to offer the appearance of change (i.e., in the "gaining of voice") that covers over the continued replication of gross disparities and the ways in which cultural representation does not extend to political decision-making power.

The final period described in Melamed's text is the neoliberal multiculturalism of the 2000s that recycles liberal multicultural language now in terms of "open society" and "economic freedoms" and "consumerist diversity." This method enacts racist "official antiracism" by assigning to all humans a neoliberal personhood premised on the belief that the market is rightfully equipped to distribute resources and manage human life. This sets the "terms, criteria, and repertoires that racialize neoliberalism's beneficiaries as worthy multicultural global citizens and its losers as doomed by their own monoculturalism, deviance, inflexibility, criminality, and other attributes deemed antisocial," in other words doomed by their own self-imposed distance from the inherent "good" of the capitalist system.[46] This set of rhetorical maneuvers enables the global monoculturalism of dominant systems to continue under the guise of supposedly "free" multiculturalism.

To explicate the effects of neoliberal multiculturalism in the university, Melamed cites Aihwa Ong's important book *Neoliberalism as Exception*. Ong identifies in the neoliberal university a

> double movement in American higher education—a shift from the national to a transnational space for producing knowledgeable subjects, and a shift from a focus on political liberalism and multicultural diversity at home to one on neoliberalism and borderless entrepreneurial subjects abroad.

The operations of higher education are one arm of the many tentacled operations of neoliberalism "used to exert a monopoly of rationality over the practices that impact its constitution," producing in turn a "differentiated citizenship" that reproduces power hierarchies on a global scale.[47]

The "official antiracisms" that Melamed identifies together work to bolster white supremacy and "naturalize the privileges of those who benefit from present socioeconomic arrangements," as they serve to "make the dispossession of those cut off from wealth and institutional power appear fair."[48] In turn, "official antiracisms have fatally limited the possibility of overcoming racism to the mechanisms of US-led global capitalism, even as they have enabled a new kind of normalizing and rationalizing violence."[49] The effects of these official antiracist orders are cumulative. All three of the periods outlined in Melamed's book are still with us, sedimented in policies and practice both within and outside of educational apparatuses. We may be living in the third era of Melamed's periodization, but the tropes associated with each of the three continue to characterize humanistic study. The title of Melamed's book dwells on the aporia of "to represent and to destroy": that "every representation, every deployment of a will to knowledge, entails destruction and assembly."[50] Thus, the nominal representation of antiracism may ultimately work against the project it names; official antiracisms are antagonistic toward a radical antiracist orientation. Melamed's book is a call to recognize how supposedly "good multicultural practice" is actually complicit in the destruction of cultural sovereignty.

Confronting universalism

What Melamed identifies are the stakes of humanistic study, as she also names our context. An additive, pluralistic model appearing in the performative language of antiracism fails to interrogate and dismantle white supremacist capitalist ableist colonialist cisheteropatriarchy, to adopt and modify bell hooks' intersectional term. These ideologies serve to sustain histories of oppression.

Pluralism traffics in the language of difference while insisting on universalism and sameness. As Kuokkanen notes:

> speaking through an epistemically different framework is too quickly interpreted as no more than a "difference." . . . It is based on the assumption that the difference resides only at the surface level, that there is a simple one-to-one correspondence between words as well as a straightforward route from "difference" to "sameness." In other words, it is assumed that there is a shared ontological, epistemic foundation below the surface differences. What is more, those occupying the privileged position of sameness can afford not to see the epistemic violence that always occurs during these translations.[51]

Noting the epistemic violences enacted by an assumed sameness, Gade uses the term "universalized pluralism" to identify the ways in which subjugation (of Islamic ways of knowing—which is the subject of Gade's book—and multiple knowledges as well) occurs "under monolithic frameworks."[52] Gade seeks to envision an environmental humanities in which

> Muslim views would not be cast as an alternative to mainstream environmentalism and its ongoing romantic-colonial heritage, but would be central to the main conversation. This might even change the subject away from explaining Islam to accepting environmentalism humanities' own need for critique with respect to the levy of nineteenth-century European perspectives.[53]

What Gade articulates for Muslim environmentalisms is also relevant to the many Indigenous and diasporic environmentalisms of the globe. A multiepistemic approach to the environmental humanities would seek to dwell in the overlapping, interconnected, yet also incommensurate and conflictual epistemologies that exist on earth, so as to enact a politics in which many worlds are possible. The many-worlded approach rebels against the universalist orientations of an extractivist, devastating colonial-capitalist march to cover all areas of the globe. It thwarts the overlay of the "we are all modern"[54] orientation of global capitalism, which rationalizes the profound dehumanization it enacts in the name of that modernity.

Mario Blaser points to pluriversal studies and political ontology as a resource for confronting this universalism to highlight the fact that, in Blaser's words, "[o]ntological conflicts . . . involve conflicting stories about 'what is there' and how they constitute realities in power-charged fields." The assumed universalist understanding "that the cultural differences that exist are between *perspectives* on one single reality 'out there' rule out the possibility of multiple ontologies and what is properly an ontological conflict (i.e., a conflict between different realities)."[55] The existence of multiple ontologies exposes the pernicious lie of universalism. Blaser calls for a pursuit of the ontological conflicts of cultural orientations. It is only with this respect for deep difference that dominant systems can be undermined.

Alternative thinking of alternatives

Affirming incommensurable difference runs counter to the act of "seeking alternatives," as Walter Mignolo explains: "if you look for alternatives you accept a point of reference instead of a set of existing options among which the decolonial enters claiming its legitimacy to sit at the table when global futures are being discussed."[56] Boaventura de Sousa Santos echoes the point: "we do not need alternatives so much as we need an alternative thinking of alternatives."[57] Santos continues, "[I]n order to capture the immense variety of critical discourses and practices and to valorize and maximize their transformative potential, an epistemological reconstruction is needed." This epistemological reconstitution would counter what Santos calls "monocultural diversity," as another way of naming normative pluralism and honor the reality that "the cultural, cognitive, social, ethnic-racial, productive, political, and religious diversity of the world is immense. . . . It is diversity with its own criteria for diversity, which, unlike monocultural diversity, turns inert simultaneity into complex contemporaneity."[58] This diversity escapes any essentialism and is what is meant by pluriversality.

In the history outlined in *Represent and Destroy*, described earlier, Melamed acknowledges how what she alternately calls "race radicalism" and "materialist antiracism" has countered hegemonic antiracism. She cites "race-based social movements that actively rejected . . . cultural normativity": among these are the diverse and non-analogous movements of Black Power, Black feminism, Chicano nationalism, Asian American civil rights, American Indian sovereignty, queer, third worldists, women of color." She continues, "Demands for third-world colleges, black and ethnic studies departments, La Raza studies, Asian American studies, and Native American studies," she writes, also "were attempts to seize the authorizing power of massive racializing institutions and the material power to produce, validate, and bind knowledge to multiracial, democratized power." She acknowledges that "Calling [these movements] all race radical risks missing antagonisms and reducing their complexity, yet it enables examination of them together in their aspect as materialist antiracists pitched against the dematerializations of official antiracists."[59] Each of these unique configurations, and the pluralities of understanding and ways of being that each represents, have in common the refusal to be thought of as alternatives to the dominant: they are instead fundamentally destabilizing of the dominant.

What would it mean to consistently ground an environmental humanities in these movements, leaving behind the white-dominated environmental humanities that is thoroughly embedded in the official movements that Melamed chronicles (i.e., racial liberalism, liberal multiculturalism, and neoliberal multiculturalism)? Racism is everywhere in the white humanities; to account for this fact is to turn the discipline inside out, dislodging centers of power.

Priscilla Ybarra charts a course for transformative decentering in the environmental humanities: In *Writing the Goodlife*, Ybarra identifies several assumptions built into white-dominated ecocriticism, recounting the field's history: the first wave characterized by the "tacit approval of settler-colonialism implicit in its. . . . enthusiasm for the pastoral and the biocentric"; the second wave including a turn to environmental justice concerns, moving to engage naturecultures and human-to-human relations and systems of exploitation, yet falling short of a truly decolonial praxis.[60] While the turn to social justice in the second wave is worthy, Ybarra argues that it is partial; there is work to be done to dismantle epistemological hierarchies and enact decolonial approaches to ecocriticism and the environmental humanities. Ybarra's history of ecocriticism aligns with the broader history articulated by Melamed, and Ybarra's argument in the book pushes toward a more radical future for the field. Ybarra's third wave is still coming into being: an anticipated future for the field that orients toward a radical decolonial politics.

Asset-based pedagogies and normative pluralism

The work of dismantling "epistemological hierarchies"—which, Ybarra notes, are characteristic of the environmental humanities—is related to several important movements in educational theory that foreground radical cultural diversity, including funds of knowledge approaches, culturally relevant and culturally sustaining pedagogies, and the recognition of community cultural wealth. These pedagogical movements affirm the cultural knowledges, values, and practices that have been devalued in white supremacist educational spaces. Together, they offer a critique of the racism embedded in conventional pedagogies.

"Funds of knowledge" is a term used in primary and secondary education scholarship beginning in the 1990s and subsequently translated to higher education by, for instance, the essay by Norma González, Ellen McIntyre, and Ann S. Rosebery's "Seeing, Believing, and Taking Action," printed in the collection *Classroom Diversity: Connecting Curriculum to Students' Lives*, and Estela Bensimon's 2007 presidential address to the Association for the Study of Higher Education. Funds of knowledge approaches expose how the understandings brought to the classroom are too often limited and inscribed within white supremacy. These conversations call educators to note how we each speak from a specific subject-position and a culturally contingent point-of-view; at the same time, these approaches call us to stretch our syllabi to draw from multiple funds of knowledge, grounded in varied cultural traditions.

Funds of knowledge approaches are one example under a broader rubric of asset-based pedagogies, which trace their roots to 1960s and 1970s critiques of deficit-based assessments of student performance. From these critiques emerged a range of asset-based pedagogical theories and affirmations of students' cultural literacies, including the April 1974 Conference on College Composition and Communication statement on "Students' Right to Their Own Language."[61]

Asset-based pedagogies have evolved significantly in the subsequent decades, further addressing the political dimensions of learning and the racism of educational systems. Asset-based pedagogies have continued the pursuit of inclusive models where all students may be affirmed. Culturally relevant pedagogy became influential with the 1995 publication of Gloria Ladson-Bilings' "Toward a Theory of Culturally Relevant Pedagogy." In this piece, Ladson-Bilings identifies three overlapping components of culturally relevant pedagogy: (1) a focus on student learning across differences, (2) an understanding that all students have culture and that it is important to activate cultural knowledges (requiring a level of cultural competence), and (3) an emphasis on sociopolitical and critical consciousness.

Following from culturally relevant approaches, culturally sustaining approaches emerge as a "generative loving critique of previous asset pedagogies."[62] Culturally sustaining approaches enable cultural continuance and the dismantling of white supremacy. As H. Samy Alim and Django Paris describe in the introduction to the anthology *Culturally Sustaining Pedagogies*, the goal is to "develop a pedagogical agenda that does not confine itself to the seemingly panoptic *White Gaze*."[63] Culturally sustaining pedagogy asks:

> Under the latest iterations of White Supremacist, capitalist, cisheteropatriarchal ideologies, systems, and practices, what knowledges must we sustain in order to overcome and survive when faced with a power that seeks to sustain itself above and beyond—and sometimes shot through—our bodies?[64]

This question is core to culturally sustaining orientations, which seek to create the conditions that counter epistemicide or the erasure and endangering of cultural knowledges.

Tara J. Yosso's community cultural wealth model has also been largely influential.[65] Also from a strength-based or asset-based perspective, Yosso demonstrates the forms of cultural capital that students from communities of color come to the university with: aspirational wealth (hopes and expectations for the future), linguistic wealth (multilingualism and other forms of versatility in communication), familial wealth (including community networks and a sense of communalism), social wealth (social networks and involvements), navigational wealth (ability to succeed within a range of environments), and resistance wealth (a continued legacy of engaging in activism and solutions-building). The naming of these forms of wealth is not meant to negate the innumerable strengths (intellectual, social, and so on) of students. Yosso's model is regularly mobilized in teaching and learning conversations, with the claim that affirming these forms of cultural capital can increase a sense of belonging and academic efficacy.

Together these conversations aim to counter white supremacist and assimilationist orientations, countering deficit-based thinking about students of color that manifest allegiance to a white-supremacist system. These theories aim to subvert, dismantle, and shove out white supremacy culture in educational spaces, as they interrupt patterns of power and privilege that are entrenched in our classroom spaces. That said, these theories can be readily misunderstood or co-opted, reduced to buzzwords in the racial liberalism of the white professoriate. These theories, while powerful, are not immune to the insidiousness of normative pluralistic models.

Rios-Aguilar and Kiyama are critical of pedagogical theorists' "efforts to relativize or 'multiculturalize' the concept of cultural capital," asserting that the idea of cultural capital must "be placed within the struggle over power."[66] Drawing from Bourdieu, Rios-Aguilar and Kiyama remind us that capital "does not exist and function except in relation to a field"—meaning that a matrix of power relations gives capital meaning. Rios-Aguilar and Kiyama point us toward the systems that reproduce these power relations, a neglected focus of many asset-based pedagogies. They note,

> most of the available scholarship in education reduces capital to what an individual possesses or lacks in relation to a field, often without explaining or identifying the field itself and how the field relates to access and activation of various forms of capital.

They implore teachers to work with students to demolish the systems that prop up the hegemonic terms of knowledge as it stands.[67] This may entail thinking outside of the terms of capital in order to resist capitalist systems.

Within higher education, the ubiquity of pedagogical aspirations going by well-circulated rubrics such as cultural relevance, culturally sustaining approaches, decolonization, and antiracism are rarely accompanied by institutional transformation.[68] Evacuated of the thoughtful educational theory that gave rise to these terms, these rubrics at times circulate to blanket over institutional oppression, reducing racism to "cultural misunderstanding or lack of information about other cultures" and flattening social justice into merely "understanding yourself and people unlike you," none of which addresses exploitation. In turn, these misrepresentations of culturally oriented pedagogical theory may produce a number of adverse effects, including the policing of forms of resistance in the name of "civility," "cross-group understanding," and "goodwill."[69] Mohanty notes the tendency to reduce "complex structural experiences of domination and resistance . . . as individual behaviours and attitudes" so that "they can be managed while carrying on business as usual."[70] Indeed, this management may serve to only strengthen the systems in place.

Kuokkanen critiques cultural exchange models that take up the language of culturally relevant practice only to strengthen control of difference.[71] She continues,

> the idea of cultural sensitivity—of being aware of other groups' cultural behaviour—only produces a "catalogue of cultural differences' and leaves systemic oppression unaddressed. When we focus on the cultural characteristics of the "other," we are in effect suggesting that the "other" is merely different, rather than oppressed.[72]

While there is a need for acknowledgment of cultural identity as a way of asserting epistemic difference against white supremacy, this assertion of self-determination is in resistance to tokenism.

Kuokkanen addresses the harms of instrumentalism, a form of tokenism that positions people "in service of" dominant orders: this includes "practices that construct indigenous people as cultural repositories or ethnographic curiosities, be it in research, in classrooms, or at public events. Instrumentalism ensures that they remain objects, not agents or subjects, of their own knowledge and epistemes."[73] We see this tendency in the "utilization" of traditional ecological knowledges in the environmental humanities under presumptions that knowledges can be taken up, snatched from the complex onto-epistemologies they exist within, and still properly understood. Traditional ecological knowledge is often constructed in the singular as if a single body of knowledge could represent all Indigenous peoples and as if this knowledge can be swiftly imported into a pre-built curriculum. Educational spaces rely on reductive models of representation and recognition, willfully ignorant of the stakes of these knowledge claims. Such education does structural violence to the complexities of subjects in relation to complex and dynamic cultural practices.

Epistemology of ignorance

Normative pluralism emerges from what Charles Mills has called an epistemology of ignorance: ignorance that is not just the *absence* of knowledge, but the *insistence* not to know.[74] Mills defines the term as produced by a Racial Contract. He writes,

> On matters related to race, the Racial Contract prescribes for its signatories an inverted epistemology, an epistemology of ignorance, a particular pattern of localized and global cognitive dysfunctions (which are psychologically and socially functional), producing the ironic outcome that whites in general be unable to understand the world they themselves have made.[75]

In other words, white racial epistemes condition people to be "aprioristically intent on denying what is before them."[76] The ignorance of white racially conditioned knowledge is in Mills' words, "best thought of as a cognitive tendency—an inclination, a doxastic disposition—which is not insuperable."[77] This pervading cognitive tendency takes many forms: insistent color-blindness, collective amnesia, apparent obliviousness, and dismissive marginalization. As Mills notes, we see the epistemology of ignorance every time racism is described as an anomaly or singular instance (or "not the norm") or every time history is narrated in eurocentric terms (centering European thought, philosophy, literature, etc.) that fail to mention that European ascendancy fundamentally relied on colonialism in many forms and that colonial and racist thinking has been structured into European cultural output. As Mills notes, there is "all the difference in the world between saying the system is basically sound despite some unfortunate racist deviations, and saying that the polity is racially structured."[78]

We see the epistemology of ignorance in "white liberalism predicated on colorless atomic individuals and a white Marxism predicated on colorless classes in struggle"[79] and in assertions of rights that see no need to point out that the terms by which rights are granted is set by some and denied to others. This "white amnesia about the actual historical record"[80] is trained, practiced and only "psychologically possible . . . [for] those who are racially privileged, for whom race is invisible precisely because the world is structured around them, whiteness as the ground against which the figures of other races . . . appear."[81]

Importantly, this ignorance is about attention—and not inattention; it is a particular kind of knowing, not an inability to know. As Alcoff notes, we can explain ignorance "not as a feature of *neglectful* epistemic practice but as a *substantive* practice in itself."[82] It is constitutive of white thought. This active ignorance is substantiated by what is attended to and how it is attended to. Elision is an active process of reducing and collapsing; it is not about what is innocently "missed." We can think of the epistemology of ignorance as a trained denial that is realized in moment-by-moment ways of knowing/doing/being that are embedded within larger systems. Countering tendencies in this onto-epistemic conduct requires resources for noticing and retraining habits of attention, a point I take up in Chapter 3.

Dorothy Wang describes how epistemologies of ignorance show up in daily practice: "If they exist at all, conversations on race," she writes, "suffer from, variously, inhibition, defensiveness, paucity of signifiers, a narrow range of possible preordained positions, caricatures of thought on all sides."[83] Claudia Rankine and Beth Loffreda, too, note how racism is commonly described only through languages of avoidance: framing racism as scandal, as existing only in past tense, or as a shock or source of sentiment. These are evasive moves that permeate multiple discursive arenas, including those associated with the university. Rankine and Loffreda reflect that for white people, "whiteness has veiled from them their own power to wound, has cut down their sympathy to a smaller size, has persuaded them that their imagination is uninflected, unfiltrated. *It has made them unknowing*."[84]

It should be clear that this ignorance is not based on a lack of knowledge; it is not about a lack of information. It is not simply that privilege has prevented a white person from having access to knowledge of racism, a simple awareness gap or blank spot. A white epistemology of ignorance is, in Wioleta Polinska's words, following Elizabeth Spelman, "a deeply entrenched epistemic resistance to know. . . . [R]acial ignorance is not a matter of simple neglect, self-deception, passivity, or accidental lack of knowledge. Rather, it is an ignorance that is actively created and managed."[85] Thus, an epistemology of ignorance is not an epistemology of absence—it is populated with fast belief. It is ignorance "actively produced for purposes of domination and exploitation" and the continuation of supremacist structures.[86]

The epistemology of ignorance produces whitewashed histories that, for example, portray "universities as having benevolent, paternalistic, inclusive relationships with African-American people both during slavery and post-emancipation while forgetting the histories of slavery and racialized exploitation that have been central for universities' foundations of ongoing existence."[87] This is one reason Boggs et al. call for an abolitionist approach to the university, as one mechanism for dismantling the "epistemological refusal of knowledge that constitutes the institution as such" that persistently normalizes "modes of forgetting behind the veil of institutional obviousness."[88] The maintenance of willfully ignorant frames is one means by which white supremacy culture rolls out its assurances and keeps divisions in place.

Kuokkanen explicates some of the ways that sanctioned ignorance (a term borrowed from Gayatri Chakravorty Spivak's *A Critique of Postcolonial Reason*)[89] is sutured into the repetitions of institutional practices. She writes of how epistemic ignorance

> is veiled in sentiments of political correctness (e.g., mainstream faculty are not permitted to teach issues pertaining to the "other"), concerns about colonialism (e.g., teaching about the "other" signifies colonization), and even "cannibalism" (e.g., the fear of "consuming" indigenous practices).[90]

In other words, forms of neglected accountability can come cloaked in the lexicon of social justice. In contrast, truly encountering radical alterity is not essentializing or co-opting and is relentlessly conscious of relations of power. The teacher does not teach "the other," but instead supports forms of relationality that insist on the presence of radical difference. To do otherwise risks the erasure that may continue what Boaventura de Sousa Santos calls "epistemicide," or the destruction of epistemologies and worldsenses that accompanies cultural genocide.

Santos elucidates ways in which epistemicide is continually carried out, limning what he calls a sociology of absences. The term sociology of absences usefully describes, as Arturo Escobar summarizes, that "what does not exist is actively produced as nonexistence or as a noncredible alternative to what exists."[91] White ignorance produces absence; it disappears ways of knowing in its hierarchies. In turn, white ignorance translates into the dismissal of treaty agreements, the continued occupation of native lands, enslavement, dehumanization, the failure to pay reparations, and to acknowledge all that is owed that can never be repaid. The regime of knowledge signified by the "epistemology of ignorance" operates as part of an ongoing colonial matrix that has enveloped the globe—although, we must remember, not in totality: resistance and survivance continues.

Epistemic justice is a project that at once confronts dominant epistemologies of ignorance, as it also grapples with the incommensurability of multiple subaltern epistemologies. This epistemological work is done in service of material transformation and the redress of systems of oppression.

On not-knowing

To dismantle modes of elision and erasure, we need to, in M. Jacqui Alexander's words, "become fluent in each other's histories." One needs to be able to recognize multiple epistemologies, while acknowledging all one does not know of them. This is a posture of *seeking*, not knowing.

The environmental humanities may be equipped to support a posture of *openness toward*, an orientation that does not resolve into what can be taken or used. In the environmental humanities, we dwell in complexities of meaning, we approach what we study from multiple theoretical angles, and we are skeptical of fixed knowledge relays and the demand to quantify. The topoi of humanistic study allow for, and even insist upon, ambiguities, for "both/and" thinking. Within humanistic study, we can exist in theoretical spaces that allow for the possibility of multiepistemic engagement. The humanities preserve ways of thinking that can counter a "discourse of certainty" (Melamed's term)[92] and universalism.

Importantly, the stance of perpetual seeking, in its commitment to not already knowing, is very different from—and, indeed, is actively opposed to—condoning what is described earlier as ignorance: it counters the sanctioned ignorance that is perpetuated in predominantly white spaces. The epistemic demeanor of seeking is aware, relational, and accountable—while an epistemology of ignorance is unyielding, removed, inert, and defiant in its stance: "I have no need to know."

"Not knowing" or, after multiple Indigenous traditions, the idea of "coming-to-know"[93] is not a passive stance; it is an active one that guards against any sense of mastering or consuming another. Paula Gunn Allen writes of Pueblo cultural practice in which "a person is expected to know no more than is necessary, sufficient and congruent with their spiritual and social place. One does not tell or inquire about matters that do not directly concern one."[94] The educational orientation that demands to know, that presumes an all-encompassing and universalizing reach, taken in by a singular knower should be called into question and no longer allowed to govern educational systems. As it oversteps, this educational orientation also tends toward objectifying knowledges: turning them into "curios, artifacts, fetishes," discussing even what is most sacred as though an "object of interest and patronization."[95]

Not knowing reflects an ecological orientation. As Deborah Bird Rose notes, "Knowledge in ecological systems is never complete, it is always changing, emerging, and fraught with uncertainty. . . . Ecological thinking takes us away from certainty." She continues, underscoring that these domains are central to Indigenous thought, "Connectivity entails *inter*dependence and brings us into domains of responsibility, accountability, proximity, ethics, and community."[96]

This posture of not-knowing is the heart of cultural humility, an approach that has been forwarded in the health professions, in diversity, equity, and inclusion (DE&I), and elsewhere. Humility entails a posture of staying close and staying in relationship, rather than achieving a resolved sense of understanding. In seeking, we refuse to settle with knowledge. As such, humility is a form of what José Medina calls a "meta-lucidity," the capacity to see the limitations of dominant ways of knowing.[97] It throws off the "excess of epistemic authority"[98] that comes from being positioned as a beneficiary of overlapping global oppressive systems.

Cultural humility was suggested as a framework by Melanie Tervalon and Jann Murray-García in a 1998 landmark essay originally intended for physicians and medical students. Proposed as a counter to competence models that assume that the empowered specialist is the "subject who knows," cultural humility as a framework has been an intervention in the hierarchy of provider–patient relations in healthcare. Rather than speaking from a position of mastery of knowledge, cultural humility invites the expert to dwell in all one does not know. A culturally humble approach has several hallmarks, including "a lifelong commitment to self-evaluation and critique, to redressing the power imbalances . . ., and to developing mutually beneficial and non-paternalistic[99] partnerships with communities on behalf of individuals and defined populations."[100] Cultural humility represents awareness of radically different worldsenses, cultural inheritances, and ways of being. The embrace of radical diversity is inclusive of neurodiversity,[101] embodied diversity, and so on. The "cultural" in cultural humility is also the embodied.

Cultural humility is actively present to difference and, in this, is relational and conscious of how continued histories come to bear upon these relationships, seeking to bring continued disparities into check.[102] The framework rejects the act of diagnosis made from the position of having "authority on" or "authority over"[103]; it is thus a tool for dismantling paternalistic systems in medicine—and, in turn, other areas such as education. Cultural humility has no "discrete endpoint" but is instead "a commitment and active engagement in a lifelong process that individuals enter into on an ongoing basis with patience, communities, colleagues, and themselves" and commitment to dismantling systems that sustain inequity.[104] Humility is not about becoming more weak or meek. It does not collapse into passivity; at the same time, it persistently resists the hierarchical structures of "knowing best" and "knowing for." As Tervalon and Murray-García describe, a culturally humble approach "relinquishes the role of the *expert* to the patient, becoming the *student* of the patient with a conviction and explicit expression of the patient's potential to be a capable and full partner in the therapeutic alliance."[105]

Cultural humility resists "mastery" as a value or aim; it instead is practice-based, in-the-moment, and processual. The call is not to essentialize cultural difference, not to become "competent" in a culture or demonstrate knowledge.

Instead, cultural humility resists the impulse to close someone or something into the "already known." It may be that only in this refusal to *already know* that someone can come to *feel known*—or feel recognized in full humanity—in a dynamic and interconnected way. In other words, by curbing a desire to approach a situation with knowledge, we become opened up to the complexity, contingency, and ongoingness of each encounter.

Importantly, cultural humility is not about content—it resists "aboutness." It does not hold content in one place. Culture is not a finite content that can be "understood." Instead, our approach to culture should be fundamentally relational, a collective engagement in who we are and where we are coming from. Making room for the dynamism of culture is an important way of sustaining it, rather than reducing it. My work of relating is not to assume knowledge or to acquire anything. Rather, a relational orientation stays fundamentally open to the "you" that I relate to. This is active and processual in dismantling hierarchy.

What Tervalon and Murray-García describe of institutionalized relationships within the medical practice also applies to the teaching practice. From Foucault, we know that educational institutions can be understood through the power relations that we read in the history of medicine. Just as Tervalon and Murray-García intervene in medical discourse and practice, their intervention also translates to educational institutions. We can use the aspiration of "developing mutually beneficial and non-paternalistic partnerships"[106] to guide classroom and research practice. The reorientation invites us to no longer think of ourselves as students and teachers, but rather each as fellow beings on this earth that is a multiverse, composed of innumerable ecosystems and ways of knowing. We encounter this earth and each other in awareness and respect for this unquantifiable relationality. Core to this orientation is a subverting of traditional models of expertise.

We do not *already know* our students, their histories, and their ways of being in the world, and we don't already know what they need. The relational process of cultural humility is equal parts learning, vulnerability, trust, willingness to be moved by another, critical interrogation of positionality, and the transformation of overlapping systems of oppression. Within the framework of cultural humility, we understand that a culturally responsive pedagogy is not a simple matter of plugging in the right readings in the right weeks on the syllabus. Epistemic humility prompts practitioners to, in Tervalon and Murray-García's words, "think consciously about their own, often ill-defined and multidimensional cultural identities and backgrounds"[107] and subject-positions within systems of inequity. We come to understand how our views are contingent, embedded in histories that have been, and continue to be, shaped by colonialism.

Positionality and cultural humility

It perhaps could, but should not, go without saying that humility and vulnerability mean differently for those who occupy different subject-positions. While the term cultural humility was proposed by two women of color as a broadly applicable framework, we come to it, and anything else, with different positionalities. Humility may feel different for those whose legitimacy has been in question by the power regimes in place and the continued legacies of ongoing colonialism and racism. Humility means differently if you've had to fight to preserve your understanding of—and right to be in—the world. Being forced into a humble posture has had the most damaging, horrifically life-upending—and devastatingly life-ending—effects for those who are dehumanized by oppressive systems.

Cultural humility calls for attention to positionality, the consideration of what it means for me to say what I say, given my identities and my context. John Johnson defines positionality as

> the ascribed value and differentials in power associated with particular social group identities in a given context and how identity constellations operate to influence how we experience the environment, both in terms of how we are treated and how we act upon others.[108]

This includes our subject-positions, social group identities, and roles (e.g., doctor, teacher) as these are structured within overlapping systems. Positionality shapes every communicative act; it conditions what can be said, how something is heard and received, and more. Attention to positionality requires an understanding of the ways that power is configured in each circumstance. To redress power imbalances requires constant grappling with how our meanings arrive and carry forward differently, depending on our social positioning. Feminist standpoint theory—with the work of Gayatri Chakravorty Spivak on sanctioned ignorance, Edward Said on positional superiority, and many others—has called for an accounting of epistemic privilege.[109]

Cultural humility is only powerful if it works against the normative. Notably, cultural humility can be a tolerance model cloaked in a new language, one more way in which racial liberals or neoliberal multiculturalists may, to return to Melamed's instructive analysis, mobilize supposed "cultural solutions" that only legitimate white privilege and consolidate class power behind a mirage of multicultural inclusiveness and equal recognition.[110] Cultural humility without economic redistribution is part of the hegemonic "official antiracism" that Melamed describes, or what Felice Blake, Paula Ioanide, and Alison Reed have called "Antiracism, Inc."[111]

Cultural humility can be part of movement building toward systemic change and regeneration if it truly undermines hierarchical relationships and if it is accompanied by far-reaching forms of material change. The risk in introducing cultural humility is that it will fall into just "another teaching about getting along with each other, or about what we ought to do so we can live happily together," which, Zenju Earthlyn Manuel writes, we do not need. The work is "not about hope for some future. It is not about diversity in the tight confines of appearance or best efforts."[112] What Manuel calls the "way of tenderness," a contemplative practice grounded in Zen Buddhism and actively attentive to the workings of race and other identity categories, "is not a method for how to behave, how to be kind, or how to transform our behaviour with each other. By this point, we have grown past formulas, techniques, and strategies that teach us how to honor life."[113] Manuel's practice suggests a cultural humility that is born out of respect for cultural sovereignty, that acknowledges the need for cultural sanctuary, and that remains deliberate in its pursuit of radical and thoroughgoing transformation.

Cultivating an epistemic demeanor

Cultural humility is an "epistemic demeanor" to bring in a term from Helen Verran.[114] Verran offers the term to, in Mario Blaser and Marisol de la Cadena's summary, "enable an ethical politics of doing difference together without any participating know-how canceling any other one."[115] Cultivating a desirable epistemic demeanor involves two attentions: (1) emergent attention, or what Verran calls *good faith*, meaning an attention "to what emerges in the here and now" and (2) attention toward what is imposed—the imposed being what Verran calls *bad will*. We learn to practice an alert criticality of "bad will" or habituated and imposed knowledge. To recognize "bad will" is to "reject the temptation, always present, to propose one's common sense to think difference."[116] Because these multiple attentions—the noticing of both the emergent (good faith) and harmful tendencies of imposition (bad will)—are sustained in a moment-by-moment unfolding, and because the risks they meet are always present, these attentions may necessitate mindfulness.

As I suggest in Chapter 3, mindfulness might be another name for this capacity for continual attention—necessary to cultural humility therefore, even as we also do well to cultivate a culturally humble orientation toward mindfulness itself, recognizing its multiple cultural roots and its varied legacy in tradition. Radical mindfulness (that does not resolve into an idea of "mind" from western enlightenment thinking) may serve as a guard against an impositional insistence on one's own "common sense," which may show up in a "routinized debunking habit,"[117] or a tendency to dismiss anything that is not within one's inherited "common sense" assumptions of truth. The "debunking habit" has such prevalence, Blaser and de la Cadena observe, that "even the possibility of questioning such a habit is met with a smirk"; that is, the habit insists upon itself by dismissing any question of it; it arrogantly rejects the encounter. In contrast, a culturally humble epistemic demeanor can be understood as the "falling to bits of the ontic and epistemic commitments" of a worldview that insists on dismissing anything outside of it.[118]

Cultural humility is an act of resistance against hegemonic regimes of truth, a pronounced turn from "knowing best" and "knowing for" that is the stance of the missionary, the colonial figure, the master, the white savior. As Paulette Regan writes in *Unsettling the Settler Within*, "the sympathetic humanitarian eye is no less a product of deeply held colonialist values, and no less authoritative in the mastery of its object than the surveying and policing eye."[119] The authoritative stance of "knowing best" and "knowing for" is pervasive in a white supremacist, colonial culture, sometimes half-concealed in the gestures of "white allyship." For example, a sympathetic white benevolence lurks behind some forms of inclusive education that serve "the colonizer whose benevolent imperialism assumes both herself or himself [*sic*] as the center of knowing and that everything can be known."[120] We find this "benevolent" orientation in understandings of higher education that seek to "remediate" students' "deficiencies," "poor performance," or "achievement gaps" while keeping in place a fundamentally racist institutional system that, by definition, is responsible for those "gaps" that are based on race. In the words of Shaun Harper, it is the institution (and the administrators and faculty that sustain dominant institutional orientations), not its students, that is in need of remediation.[121] The institution, and those who benefit from it, must account for the long legacy of reproducing racist stratification. Yet rarely is this responsibility acknowledged; instead, "deficit model" thinking prevails whenever students' "lack" (of education, skills, ability, commitment, etc.) is what is blamed for the disparities that exist in education, in place of considering what the university lacks in terms of representation, cultural relevance, equitable policies, humanizing and affirming practices, or otherwise.

In coming to consciousness of radical difference, there is no goal of "taking on." I can't simply step into a new epistemology, but I can adopt an active stance of "not knowing"—meaning that I let go of a desire to appropriate, fold into my existing schemas, or make my own. We can think of this orientation as very different from "a liberal desire to understand everyone."[122] What we might call "pluriversal humility" and "epistemic humility" provides the means of tracing a frame as a frame—in other words, cultural humility enables us to see that our understanding is never neutral, universal, or absolute. Pluriversal humility resists easy notions of understanding. By "not knowing," I pursue the art of seeking, grappling with the complexities of what it means to stay present to an encounter, becoming more keenly aware without resting back in the status of "already knowing," as I also am guided by my commitments to socially just transformation.

Walking, we ask questions

In her memoir *Unbowed*, Wangari Maathai uses an fluvial metaphor—"tributaries of knowledge"—to describe the importance of multiple cultural streams of knowing. We draw maps to better understand their complex cultural ecologies. These traditional waters flow; they're not compartmentalized or frozen in time. At the same time, it is important to know the differences between one body of water and another, even as they flow together as part of an interdependent earth system.

This "multiple tributaries" approach transcends interdisciplinarity—specifically, the forms of interdisciplinarity that are celebrated for epistemological versatility even as they stay within white-centric ways of knowing. This is the topic of the next chapter.

Boaventura de Sousa Santos writes:

> Before the infinite, no arrogance, only humility, is possible. Humility does not mean negativity or skepticism. . . . The fact that it is not possible to reach the truth with accuracy does not release us from searching for it. Quite the opposite, what lies beyond limits (the truth) rules what is possible and demandable within the limits.[123]

We need to go further toward a relational accountability that emphasizes reciprocity and interconnection—going further into work that is infinite.

As the Zapatistas are well-known for the saying, *Preguntando, caminamos,* or "Walking, we ask questions."[124] The questions, the unknowing will always remain, but we move in relation to one another, as best we can, moving toward a fuller acknowledgment of the worldsenses we live in and the worlds that are possible when the dominant is dismantled. This is not a sentimentalized notion.

Notes

1 Charles Mills, "White Ignorance," *Race and Epistemologies of Ignorance*, edited by Shannon Sullivan and Nancy Tuana (SUNY P, 2007) 32.
2 José Medina, *The Epistemology of Resistance: Gender and Racial Oppression, Epistemic Injustice, and Resistant Imaginations* (Oxford UP, 2013) 3, 28.
3 Gayatri Chakravorty Spivak, *Can the Subaltern Speak?: Reflections on the History of an Idea* (Columbia UP, 2010).
4 Indigenous Principles of Just Transition. Indigenous Environmental Network. 2018, www.ienearth.org/wp-content/uploads/2018/08/PrinciplesJustTransition-BW.pdf.
5 Charles Mills, *The Racial Contract* (Cornell UP, 1997) 1.
6 Mills, *Racial Contract*, 4.
7 Nora Berenstain, "White Feminist Gaslighting," *Hypatia* 35 (2020): 733–758.
8 Following the publication of *Epistemic Injustice*, Fricker has been in conversation with Mills. See, for example, Miranda Fricker, "How Is Hermeneutical Injustice Related to 'White Ignorance'?" *Social Epistemology Review and Reply Collective* 2.8 (2013): 49–53. See also: Charles Mills, "White Ignorance and Hermeneutical Injustice: A Comment on Medina and Fricker," *Social Epistemology Review and Reply Collective* 3.1 (2013): 38–43, http://wp.me/p1Bfg0-1ce.
9 Vandana Shiva, *Monocultures of the Mind* (Zed, 1993) 7.
10 Boaventura de Sousa Santos, *Epistemologies of the South: Justice Against Epistemicide*, 2nd edition (Routledge, 2016) 13; Shiva, *Monocultures of the Mind*, 14.
11 Shiva's oeuvre represents a sharp critique of neoliberalism's globalizing forces, and she notes the ontological implications. Escobar also takes up this lens: "scratch a globalization discourse . . . you will find *Homo Oeconomicus* at large." By this, he refers to the neoliberal regimes that make individuals into "miniature capitalist clones" through the operations of microfinance, modernization, and rural development. Built into these neoliberal regimes are "assumptions of linear rationality as the default mode of thinking of those wanting to make it in a competitive world, a view of nature as a resource to be extracted at any cost" (Escobar 82).
12 Shiva, *Monocultures of the Mind*, 12.
13 Mills, *Racial Contract*, 18.
14 Mills, *Racial Contract*, 19.
15 Zoe Todd, "Indigenizing the Anthropocene," *Art in the Anthropocene: Encounters Among Aesthetics, Politics, Environments, and Epistemologies* (Open Humanities P, 2015) 244.

16 Heather Davis and Zoe Todd, "On the Importance of a Date, or Decolonizing the Anthropocene," *ACME: An International Journal for Critical Geographers* 16.4 (2017): 763.
17 Davis and Todd, "On the Importance of a Date," 763.
18 Todd "Indigenizing the Anthropocene," 244.
19 Davis and Todd, "On the Importance of a Date," 761.
20 Davis and Todd, "On the Importance of a Date," 771.
21 Capitalocene (Haraway and Kenney, 2015; Malm 2013; Moore 2015), Eurocene (Grove 2016), White Supremacy Scene (Mirzoeff 2016), Chthulucene (Haraway 2016). Citations are included in Davis and Todd, "On the Importance of a Date," 763. Donna Haraway and Martha Kenney, "Anthropocene, Capitalocene, Chthulucene," *Art in the Anthropocene: Encounters Among Aesthetics, Politics, Environments, and Epistemologies*, edited by Heather Davis and Etienne Turpin (Open Humanities P, 2015) 255–270. Andreas Malm, "The Origins of Fossil Capital: From Water to Steam in the British Cotton Industry," *Historical Materialism* 21.1 (2013): 15–68. Jason Moore, *Capitalism in the Web of Life: Ecology and the Accumulation of Capital* (Verso, 2015). Jairus Grove, "Response to Jedediah Purdy," *Boston Review* (11 January 2016). Nicholas Mirzoeff, "It's Not the Anthropocene, It's the White Supremacy Scene, Or, The Geological Color Line," *After Extinction*, edited by Richard Grusin (U of Minnesota P, 2016). Donna Haraway, *Staying with the Trouble: Making Kin in the Chthulucene* (Duke UP, 2016). See also Anna Tsing, "Unruly Edges: Mushrooms as Companion Species," *Environmental Humanities* 1.1 (2012): 141–154.
22 Anna Gade, *Muslim Environmentalisms* (Columbia UP, 2019) 253.
23 Gade, *Muslim Environmentalism*, 2.
24 Shiva, *Monocultures of the Mind*, 9.
25 Shiva, *Monocultures of the Mind*, 65.
26 "Ontological designs" is Escobar's term, influenced by Terry Winograd and Fernando Flores's *Understanding Computers and Cognition*.
27 Alexander, *Pedagogies of Crossing*, 6.
28 DuCille, *Skin Trade*, 123. Quoted in Alexander, *Pedagogies of Crossing*.
29 The term "normative pluralism" appears in many sources. In *Teachers as Intellectuals: Toward a Critical Pedagogy of Learning*, Henry Giroux critiques "the pedagogy of normative pluralism" (95).
30 Chandra Talpade Mohanty, *Feminism Without Borders: Decolonizing Theory, Practicing Solidarity* (Duke UP, 2003) 204.
31 Mohanty, *Feminism Without Borders*, 211.
32 Mohanty, *Feminism Without Borders*, 188.
33 María Lugones, "Hablando cara a cara / Speaking Face to Face: An Exploration of Ethnocentric Racism," *Making Face, Making Soul/Haciendo Caras: Creative and Critical Perspectives by Feminists of Color*, edited by Gloria Anzaldúa (Aunt Lute Books, 1995) 51.
34 AnaLouise Keating, *Transformation Now!: Toward a Post-Oppositional Politics of Change* (U of Illinois P, 2013) 68.
35 Normative pluralism's "benign coexistence" has an additional problem: What AnaLouise Keating calls "separatist multiculturalism," normative pluralism tends to fix identities into neat categories, taxonomizing and tokenizing while failing to grapple with claims to land, culture, knowledge, self-determination, and so on.
36 Rauna Kuokkanen, *Reshaping the University: Responsibility, Indigenous Epistemes, and the Logic of the Gift* (U of British Columbia P, 2007) 59.
37 Kuokkanen, *Reshaping the University*, 79.
38 I wish to thank Andrea Delgado for introducing me to this source and for facilitated discussions of it.
39 Melamed's text is aligned with Chandra Talpade Mohanty's *Feminism without Borders*, which takes up the critique of "the generation of discourses of diversity and pluralism grounded in an apolitical, often individualized identity politics . . . that is different from recognizing the significance of the politics of identity" (Mohanty 107).

40 Jodi Melamed, *Represent and Destroy: Rationalizing Violence in the New Racial Capitalism* (U of Minnesota P, 2011) 9.
41 Melamed, *Represent and Destroy*, xiv.
42 bell hooks, *Teaching Community: A Pedagogy of Hope* (Routledge, 2003) 47.
43 Melamed, *Represent and Destroy*, xiv–xv.
44 Melamed, *Represent and Destroy*, 37, 34.
45 Melamed, *Represent and Destroy*, 97.
46 Melamed, *Represent and Destroy*, 39, 138.
47 Aihwa Ong, *Neoliberalism as Exception: Mutations in Citizenship and Sovereignty* (Duke UP, 2006), 140, 41.
48 Melamed, *Represent and Destroy*, 10, 2.
49 Melamed, *Represent and Destroy*, xi.
50 Melamed, *Represent and Destroy*, 230.
51 Kuokkanen, *Reshaping the University*, 76.
52 Anna Gade, *Muslim Environmentalisms* (Columbia UP, 2019) 248.
53 Gade, *Muslim Environmentalisms*, 249.
54 Mario Blaser, "Ontological Conflicts and the Stories of Peoples in Spite of Europe," *Current Anthropology* 54.5 (2013): 548.
55 Blaser, "Ontological Conflicts," 548.
56 Walter D. Mignolo, *The Darker Side of Western Modernity: Global Futures, Decolonial Options* (Duke UP, 2011), xxviii.
57 Santos, *Epistemologies of the South*, 42.
58 Santos, *Epistemologies of the South*, 13.
59 Melamed, *Represent and Destroy*, 26, 29, 48.
60 Priscilla Solis Ybarra, *Writing the Goodlife: Mexican American Literature and the Environment* (U of Arizona P, 2016) 10, 13.
61 Importantly, this statement has been recently revised. See "This Ain't Another Statement! This Is a DEMAND for Black Linguistic Justice!," *Conference on College Composition and Communication* (July 2020), https://cccc.ncte.org/cccc/demand-for-black-linguistic-justice.
62 H. Samy Alim and Django Paris, *Culturally Sustaining Pedagogies: Teaching and Learning for Justice in a Changing World* (Teachers College, 2017) 4.
63 Alim and Paris, *Culturally Sustaining Pedagogies*, 2.
64 Alim and Paris, *Culturally Sustaining Pedagogies*, 14.
65 Tara J. Yosso, "Whose Culture Has Capital? A Critical Race Theory Discussion of Community Cultural Wealth," *Race Ethnicity and Education* 8.1 (2005): 69–91.
66 Cecilia Rios-Aguilar and Judy Marquez Kiyama, *Funds of Knowledge in Higher Education, Honoring Students' Cultural Experiences and Resources as Strengths* (Routledge, 2018) 12.
67 Rios-Aguilar and Kiyama, *Funds of Knowledge*, 14, 13.
68 See Gina Ann Garcia, *Becoming Hispanic-Serving Institutions: Opportunities for Colleges and Universities* (Johns Hopkins UP, 2019).
69 Mohanty, *Feminism Without Borders*, 211.
70 Mohanty, *Feminism Without Borders*, 210.
71 Kuokkanen, *Reshaping the University*, 99.
72 Kuokkanen, *Reshaping the University*, 109.
73 Kuokkanen, *Reshaping the University*, 73.
74 This will to know and not know is related to the idea of a "white racial frame," which has been popularized by Joe R. Feagin's book of that title; to truly call the white racial frame into question entails a meta-awareness of how white epistemologies show up, how they stick, and what they foreclose. Robin DiAngelo defines the white racial frame as "the deeply internalized racist framework through which whites make racial meaning. This framework includes images, interpretations, perceptions, evaluations, emotions, and actions that position whites as superior and that are passed down and reinforced throughout society" (121).

75 Mills, *Racial Contract*, 18.
76 Mills, "White Ignorance," 27.
77 Mills, "White Ignorance," 23.
78 Mills, *Racial Contract*, 124.
79 Mills, *Racial Contract*, 110–111.
80 Mills, *Racial Contract*, 102.
81 Mills, *Racial Contract*, 76.
82 Linda Martín Alcoff, "Epistemologies of Ignorance: Three Types," *Race and Epistemologies of Ignorance*, edited by Shannon Sullivan and Nancy Tuana (SUNY P, 2007) 39.
83 Dorothy J. Wang, *Thinking Its Presence: Form, Race, and Subjectivity in Contemporary Asian American Poetry* (Stanford UP, 2014) 16.
84 Claudia Rankine and Beth Loffreda, *The Racial Imaginary: Writers on Race in the Life of the Mind* (Fence, 2015) 19.
85 Wioleta Polinska, "Mindfulness Meditation as a Remedy to 'White Ignorance' and Its Consequences," *Buddhist-Christian Studies* 38 (2018): 328.
86 Shannon Sullivan and Nancy Tuana, *Race and Epistemologies of Ignorance* (SUNY P, 2007) 1.
87 Abigail Boggs, Eli Meyerhoff, Nick Mitchell, and Zach Schwartz-Weinstein, *Abolitionist University Studies: An Invitation* (2019) 12, https://abolition.university/invitation/.
88 Boggs et al., *Abolitionist University*, 1.
89 Sanctioned ignorance is "Ignorance That They Are Allowed to Possess" (Spivak, *The Postcolonial Critic*, 60). See also *A Critique of Postcolonial Reason: Toward a History of the Vanishing Present* (Harvard UP, 1999).
90 Kuokkanen, *Reshaping the University*, 67.
91 Arturo Escobar, *Pluriversal Politics: The Real and the Possible* (Duke UP, 2020) 69.
92 Melamed, *Represent and Destroy*, xi.
93 Gregory Cajete notes that there is no word for education in most indigenous languages. Rather, education is best described as "coming-to-know" (Cajete 80, quoted in Rendón 17).
94 Paula Gunn Allen, "Special Problems in Teaching Leslie Marmon Silko's *Ceremony*," *American Indian Quarterly* 14.4 (1990): 380.
95 Allen, "Special Problems in Teaching," 383.
96 Deborah Bird Rose, "Connectivity Thinking, Animism, and the Pursuit of Liveliness," *Educational Theory* 67.4 (2017): 495.
97 Medina, *Epistemologies of Resistance*, 47.
98 Medina, *Epistemologies of Resistance*, 57.
99 Countering this paternalism is deep work. It entails uprooting the "Western arrogance of feeling that it has everything to teach others and nothing to learn from them," which is profoundly unjust, a point made by Jasmine Syedullah in *Radical Dharma* (84).
100 Melanie Tervalon and Jann Murray-García, "Cultural Humility Versus Cultural Competence: A Critical Distinction in Defining Physician Training Outcomes in Multicultural Education," *Journal of Health Care for the Poor and Underserved* 9.2 (May 1998) 123.
101 See Melanie Yergeau, *Authoring Autism: On Rhetoric and Neurological Queerness* (Duke UP, 2018).
102 Miguel E. Gallardo writes in the edited collection *Developing Cultural Humility: Embracing Race, Privilege, and Power* (2014): "Cultural humility has been defined as a lifelong process of self-reflection, self-critique, continual assessment of power imbalances, and the development of mutually respectful relationships and partnerships." Cultural humility means researching one's own identities and the histories that inform them, excavating privilege, and identifying responsibilities in a world that is characterized by inequity.
103 Thanks to Cheryl Johnson, Edelmira Reynoso, and Meridith Oram for this reference.
104 Tervalon and Murray-García, "Cultural Humility Versus Cultural Competence," 118.
105 Tervalon and Murray-García, "Cultural Humility Versus Cultural Competence," 121.
106 Tervalon and Murray-García, "Cultural Humility Versus Cultural Competence," 123.
107 Tervalon and Murray-García, "Cultural Humility Versus Cultural Competence," 120.
108 Johnson, John. Personal correspondence. 2020.

109 Gayatri Chakravorty Spivak, *An Aesthetic Education in the Era of Globalization* (Harvard UP, 2012). Said, Edward W. Said, *Orientalism* (Vintage, 1979).
110 Melamed, *Represent and Destroy*, 109.
111 Felice Blake, Paula Ioanide, and Alison Reed, eds. *Antiracism Inc.: Why the Way We Talk About Racial Justice Matters* (Punctum, 2019).
112 Zenju Earthlyn Manuel, *The Way of Tenderness: Awakening through Race, Sexuality, and Gender* (Wisdom, 2015) 10.
113 Manuel, *The Way of Tenderness*, 32.
114 Helen Verran, "The Politics of Working Cosmologies Together While Keeping Them Separate," *A World of Many Worlds*, edited by Marisol de la Cadena and Mario Blaser (Duke UP, 2018).
115 de la Cadena and Blaser, *A World of Many Worlds*, 10.
116 As summarized by de la Cadena and Blaser, *A World of Many Worlds*, 11.
117 de la Cadena and Blaser, *A World of Many Worlds*, 14.
118 Verran, "Politics of Working Cosmologies," 115.
119 Paulette Regan, *Unsettling the Settler Within: Indian Residential Schools, Truth Telling, and Reconciliation in Canada* (U of British Columbia P, 2010) 45.
120 Alison Jones and Kuni Jenkins, "Rethinking Collaboration: Working the Indigene-Colonizer Hyphen," *Handbook of Critical and Indigenous Methodologies*, edited by Norman K. Denzin, Yvonna S. Lincoln, and Linda Tuhiwai Smith (Sage, 2008) 24.
121 Shaun Harper, *Keynote Address, California State University GI205 Initiative Symposium*, 12 October 2017, https://www2.calstate.edu/csu-system/why-the-csu-matters/graduation-initiative-2025/symposium/2017.
122 Blaser, "Ontological Conflicts," 559.
123 Santos, *Epistemologies of the South*, 110.
124 Mariana Mora asked of a Tseltal member of the Zapatista autonomous government named Mauricio about the significance of a particular march that did not include spoken or written political discourse. Maurcio responded, "To us, speaking is done by walking. And speaking, which is also our walking, can be done in silence. Our word isn't done just by a few, it is in all of us, in all communities, not just the authorities. That is what autonomy is, and that is how we propose to move history." Quoted in Mariana Mora, *Kuxlejar Politics: Indigenous Autonomy, Race, and Decolonizing Research in Zapatista Communities* (U of Texas P, 2017) 3.

Works cited

Alaimo, Stacy. *Bodily Natures: Science, Environment, and the Material Self.* Indiana UP, 2010.

Alcoff, Linda Martín. "Epistemologies of Ignorance: Three Types." *Race and Epistemologies of Ignorance*, edited by Shannon Sullivan and Nancy Tuana. SUNY Press, 2007.

Alexander, M. Jacqui. *Pedagogies of Crossing: Meditations on Feminism, Sexual Politics, Memory, and the Sacred.* Duke UP, 2006.

Alim, H. Samy, and Django Paris. *Culturally Sustaining Pedagogies: Teaching and Learning for Justice in a Changing World.* Teachers College, 2017.

Allen, Paula Gunn. "Special Problems in Teaching Leslie Marmon Silko's *Ceremony*." *American Indian Quarterly* 14.4 (1990): 379–386.

Baker, Shalanda. *Revolutionary Power: An Activist's Guide to the Energy Transition.* Island Press, 2021.

Blaser, Mario. "Ontological Conflicts and the Stories of Peoples in Spite of Europe." *Current Anthropology* 54.5 (2013): 547–568.

Blaser, Mario, and Marisol de la Cadena. "Introduction: Pluriverse: Proposals for a World of Many Worlds." *A World of Many Worlds*, edited by Marisol de la Cadena and Mario Blaser. Duke UP, 2018.

Boggs, Abigail, Eli Meyerhoff, Nick Mitchell, and Zach Schwartz-Weinstein. *Abolitionist University Studies: An Invitation*, 2019. https://abolition.university/invitation/

Davis, Heather, and Zoe Todd. "On the Importance of a Date, or Decolonizing the Anthropocene." *ACME: An International Journal for Critical Geographers* 16.4 (2017): 761–780.

Escobar, Arturo. "Transition Discourses." *Constructing the Pluriverse: The Geopolitics of Knowledge*, edited by Bernd Reiter. Duke UP, 2018.

Escobar, Arturo. *Pluriversal Politics: The Real and the Possible*. Duke UP, 2020.

Gade, Anna. *Muslim Environmentalisms*. Columbia UP, 2019.

Jones, Alison, and Kuni Jenkins. "Rethinking Collaboration: Working the Indigene-Colonizer Hyphen." *Handbook of Critical and Indigenous Methodologies*, edited by Norman K. Denzin, Yvonna S. Lincoln, and Linda Tuhiwai Smith. Sage, 2008, pp. 471–486.

Keating, AnaLouise. *Transformation Now!: Toward a Post-Oppositional Politics of Change*. U of Illinois P, 2013.

Kuokkanen, Rauna. *Reshaping the University: Responsibility, Indigenous Epistemes, and the Logic of the Gift*. U of British Columbia P, 2007.

Lugones, María. "Hablando cara a cara/Speaking Face to Face: An Exploration of Ethnocentric Racism." *Making Face, Making Soul/Haciendo Caras: Creative and Critical Perspectives by Feminists of Color*, edited by Gloria Anzaldúa. Aunt Lute Books, 1995, pp. 46–55.

Maathai, Wangari. *Unbowed: A Memoir*. Reprint edition. Anchor, 2007.

Manuel, Zenju Earthlyn. *The Way of Tenderness: Awakening through Race, Sexuality, and Gender*. Wisdom, 2015.

McKittrick, Katherine. *Dear Science and Other Stories*. Duke UP, 2020.

Medina, José. *The Epistemology of Resistance: Gender and Racial Oppression, Epistemic Injustice, and Resistant Imaginations*. Oxford UP, 2013.

Melamed, Jodi. *Represent and Destroy: Rationalizing Violence in the New Racial Capitalism*. U of Minnesota P, 2011.

Mignolo, Walter. "Epistemic Disobedience, Independent Thought, and Decolonial Freedom." *Theory, Culture, and Society* 26.7–8 (2009): 159–181.

Mignolo, Walter. *The Darker Side of Western Modernity: Global Futures, Decolonial Options*. Duke UP, 2011.

Mills, Charles W. *The Racial Contract*. Cornell UP, 1997.

Mills, Charles W. "White Ignorance." *Race and Epistemologies of Ignorance*, edited by Shannon Sullivan and Nancy Tuana. SUNY Press, 2007, pp. 13–38.

Mohanty, Chandra Talpade. *Feminism Without Borders: Decolonizing Theory, Practicing Solidarity*. Duke UP, 2003.

Moll, Luis C., Cathy Amanti, and Norma González. *Funds of Knowledge: Theorizing Practice in Households, Communities, and Classrooms*. Erlbaum Associates, 2005.

Ong, Aihwa. *Neoliberalism as Exception: Mutations in Citizenship and Sovereignty*. Duke UP, 2006.

Polinska, Wioleta. "Mindfulness Meditation as a Remedy to 'White Ignorance' and Its Consequences." *Buddhist-Christian Studies* 38 (2018): 325–341.

Rankine, Claudia, and Beth Loffreda. *The Racial Imaginary: Writers on Race in the Life of the Mind*. Fence, 2015.

Regan, Paulette. *Unsettling the Settler Within: Indian Residential Schools, Truth Telling, and Reconciliation in Canada*. U of British Columbia P, 2010.

Rose, Deborah Bird. "Connectivity Thinking, Animism, and the Pursuit of Liveliness." *Educational Theory* 67.4 (2017): 491–508.

Santos, Boaventura de Sousa. *Epistemologies of the South: Justice Against Epistemicide*. Routledge, 2014.

Shiva, Vandana. *Monocultures of the Mind*. Zed, 1993.

Spivak, Gayatri Chakravorty. *Can the Subaltern Speak?: Reflections on the History of an Idea.* Columbia UP, 2010.
Sullivan, Shannon, and Nancy Tuana. "Introduction." *Race and Epistemologies of Ignorance*, edited by Shannon Sullivan and Nancy Tuana. SUNY Press, 2007.
Tervalon, Melanie, and Jann Murray-García. "Cultural Humility versus Cultural Competence: A Critical Distinction in Defining Physician Training Outcomes in Multicultural Education." *Journal of Health Care for the Poor and Underserved* 9.2 (May 1998): 117–125.
Todd, Zoe. "Indigenizing the Anthropocene." *Art in the Anthropocene: Encounters Among Aesthetics, Politics, Environments, and Epistemologies.* Open Humanities Press, 2015, pp. 241–254.
Trinh T. Minh-Ha. *When the Moon Waxes Red: Representation, Gender, and Cultural Politics.* Routledge, 1991.
Verran, Helen. "The Politics of Working Cosmologies Together While Keeping Them Separate." *A World of Many Worlds*, edited by Marisol de la Cadena and Mario Blaser. Duke UP, 2018.
Wang, Dorothy J. *Thinking Its Presence: Form, Race, and Subjectivity in Contemporary Asian American Poetry.* Stanford UP, 2014.
Ybarra, Priscilla Solis. *Writing the Goodlife: Mexican American Literature and the Environment.* U of Arizona P, 2016.
Yosso, Tara J. "Whose Culture Has Capital? A Critical Race Theory Discussion of Community Cultural Wealth." *Race Ethnicity and Education* 8.1 (2005): 69–91.

2

WHAT IS EPISTEMIC JUSTICE?

In historically white institutions, everything needs to be rethought. From how one teaches and what is taught to the conventions of academic spaces of learning, administration, and governance—every aspect of our work risks reproducing white supremacy and overlapping systems of oppression; indeed, that is what the university was built to do—to reinforce and sustain these systems. Universities have coexisted with, and benefitted from, slavery, boarding schools, internment camps, imperialist wars, and border walls, and that legacy is installed in institutional spaces. As Chandra Mohanty notes, the university has never been "a democratic public space" or "sanctuary for non-repression," as it may at times be upheld to be.[1] The university continues its work in reproducing social hierarchies and exacerbating wealth disparities, as it also reproduces epistemic injustice.[2]

Within the academy, intersectional racism is insidious, and it permeates the system's daily reproductions. Kevin Quashie names it in these terms: "racism is a profound phenomenon—it has material and psychological impact, and it is institutionally and individually unrelenting."[3] Academic work, and the value systems by which it is appraised, are heavy with perpetuated histories of racism, colonialism, and forms of exclusion and oppression. Knowledge practices have been made by these histories. The western disciplines that have been institutionalized and reproduced serve to further entrench existing hierarchies. What does it mean to trouble that within which we, to varying degrees, are embedded? Certainly, it cannot mean that those who are most inscribed in these disciplines are those most able to offer paths forward. Those who assume our belonging, those who enter the university space most unreservedly—willingly accepting its rules of knowledge production and contribution—are those who are most at risk of reproducing its harms. The cultural messages we have received in this privileged group may make us prone to thinking of ourselves as saviors, but we do not save; the world we have built for ourselves has done irreparable harm and has produced countless losses. Our role is

DOI: 10.4324/9781003037439-3

to leverage and give up our privilege as we can in support of the transformation that is collectively envisioned and coalitionally enacted.

The transformation entails the dismantling of white supremacist cultural practices. And this, redressing the consequences of the white-dominated practice, is not a matter of making minor adjustments—trimming a corner here, or making a gesture there. As Reverend angel Kyodo williams notes in a co-authored text, "the superiority of whiteness play[s] out in everyday interactions as unconscious bias, microaggressions, and a tendency to exert control over cultural norms and space." White people "find a never-ending litany of excuses to maintain power over rather than power with—to dominate."[4] When academic disciplines are maintained to reinforce white knowledges, practices, and privilege, what is needed is a counter in the form of radical imagination and the continuance and revitalization of lifeways and knowledges that have withstood colonial encroachment.

Transformative change is just that: transformative, all-encompassing. As Fanon said, it will not go unnoticed. It will not come about by a simple set of trainings; it will not be brought forward by a battle of arguments or by a single panacea. Racism cannot be fully accounted for in the administrative-managerial terms of bias prevention training. Our assumptions are tied to the epistemological; they are not merely ephemeral and perceptual, and they represent one of many dimensions of how oppression works through us. If bias is about deterministic configurations of favorability, bias structures our world. It is what enables the dominant to remain dominant. As Laura Rendón notes,

> The dominant belief system is powerful, entrenched, validated, and constantly rewarded by the social structure that created it—so much so that even when we begin to see that some of the agreements in the belief system are flawed and in need for change, we find it very difficult to challenge them.[5]

Rendón demonstrates some of the dimensions of dominance and how it wins assent in privileging particular constructs of rational knowing, binaristic thinking that is based on colonial models of division, and fundamentally objectifying orientations (e.g., the learner is separate from what is learned).[6] These dimensions of a dominant belief system structure what can be thought, said, and known.

To call into question the enforcement of legibility and livability is work that extends far beyond the classroom, although the classroom is one space where these operations of power are concentrated. To dismantle these systems as they play out in the classroom requires a radical reimagining of every relational line in the space: between teacher and student, student and student, student and self, and how each relationship is positioned and held in the space. These relationships have knowledge practices embedded within them (i.e., whose knowledges are authorized and valued in a given situation). To rethink knowledge systems is also to rethink knowledge relationships: to address not only what is favored or privileged but also what is denied. As Eve Tuck and K. Wayne Yang note, "the academy as an apparatus of settler colonial knowledge already domesticates, denies, and dominates other forms of knowledge. It too refuses. It sets limits, but disguises itself as limitless."[7] The knowledge practices of the academy become naturalized and understood as just "what is" and their constitutive exclusions are rarely mentioned: for example, the knowledges that are embedded in Indigenous languages and lifeways are elided in academic spaces which have inherited the culturally genocidal orientations of their colonial roots. Within these spaces, an appeal to "how it's done" (as in "that's just not how it's done" or "that's not what we do") should be read as a warning sign that "it" needs to be rethought, reimagined. May we, as a field, be alert to this signal.

The teaching profession has been construed as that which gives students access to disciplinary conversations as they stand. In this conception, the teacher stands at the door granting entrance to an established field and (often as an afterthought or a fraught consideration) a profession. That this is a fundamentally assimilationist orientation is rarely noticed or questioned. Our learning outcomes stay fixed in place: *this is what it means to do what we do and therefore this is what we teach and what our students must learn, no matter who they are or where they come from.*

Disciplinary conventions in a white academy—the "how it's done" of our thought life—have been made to serve white epistemologies. Those conversations that refuse white epistemologies—i.e., conversations taking place in ethnic studies, Indigenous studies, feminist of color theory, are those that, in Angela Davis' words, "refuse to subordinate knowledge to disciplinary thinking."[8] That is, these orientations resist the epistemological roots of disciplinarity as a construct, as it is an outgrowth of a "categorize and control" orientation. (That interdisciplinarity is a marked effort in our context demonstrates the extensiveness of this command; interdisciplinarity transgresses the managerial divisions that govern academic life and thus it is troublesome to the institutional apparatus. Interdisciplinary or transdisciplinarity is *accommodated*, but only partially—nominally championed in academic forums, but only to the point that it does not disrupt established structures.) With this insistence on division, the organization of knowledge represented by disciplinarity in the academy is fully entrenched in its colonial roots. Philosophy is separated from land-based disciplines. Forms of study are partitioned in ways that negate fundamental relationality. Not only is this arrangement incompatible with many examples of Indigenous thought and knowledges of communities around the globe, but it also serves to contain and shrink these knowledge systems, compartmentalizing them or relegating them to the margins of the reproduced organization of knowledge.[9]

Katherine McKittrick, in *Dear Science*, brilliantly demonstrates the relentlessness of disciplinary categorization in passages that also reveal the absurdity of the will to classify, rank, and divide a world of immeasurable interrelatedness. McKittrick identifies the will to canon-formation that is always behind disciplinarity[10]: the maintenance of the white supremacist "best that has been thought and said," as Matthew Arnold made manifest. This will-to-canonize and hierarchize the "best" and legitimate for those who have the power to make this designation further perpetuates the ways in which knowledges are disappeared. The hierarchization of knowledge is at stake in every decision made regarding resource allocation and who is brought to the decision-making table, what ways of knowing and ways of being are represented there, and who dictates the terms of these conversations.

Those orientations that exceed disciplinarity—ethnic studies, Indigenous studies, transnational feminisms, queer studies, and disability studies—may be those most likely to be disenfranchised by the academic-industrial complex. It is no coincidence that these fields are also those most able to redress "the legacies of race-making in our culture," to adopt Rhonda Magee's phrase, and other forms of stratification and oppression.[11] The thought-work represented by these fields also moves toward what can enable the thriving of all species on the planet, foregrounding healing justice, Indigenous science, decolonization, antiracist and anti-Imperialist practices, and movement-building. The manifold ways of knowing represented by these degree programs and departments should lead intellectual work in the environmental humanities and the university at large, as these ways of knowing also transcend and transform academic practice.

Importantly, these fields carry forward the vibrant legacy of student protest movements. Jodi Melamed recounts how the Black, Chicano, and ethnic studies movements on campuses worked "to transform US universities . . . so that instead of indoctrinating students into white culture and the status quo, they might radicalize them and come to work in the interest of the working class and communities of color." The student protest movements underscored the "importance of working simultaneously inside and outside universities" and of working in "the communities, epistemes, and cultures that neoliberal-multiculturalism formalism cannot transact." Melamed prioritizes the activism that cannot be incorporated into neoliberalism and "the day-to-day practices of resistance" that politicize neoliberalism's "conditions of existence."[12] With the enduring energy of these movements, the university is a continued site of counterhegemonic potential. Staying close to these sites of knowledge-production and action is how the environmental humanities can fulfill its disciplinary intention of intervening in the centuries-old relentless onslaught of ecosocial harms.

As the presence of the student protest movements attests, the academy is a contradictory place. It is a space where insurgency exists alongside entrenched knowledge structures and the disciplining mechanisms that maintain them. The university, Mohanty notes, is a place "where knowledges are colonized but also contested—a place that engenders student mobilizations and progressive movements of various kinds."[13] In the environmental humanities, what would it mean to make the promise of these student movements central to our work? How can we push forward with their momentum, throwing off inherited practices?

Part of the work entails dismantling—not just decentering—white environmentalism that rests upon racist, colonialist logics. In *The Ecological Other*, Sarah Jaquette Ray notes how the environmental movement is rooted in a politics of disgust against various communities that reinforces existing hierarchies in terms of who counts as ecologically worthy and who is ecologically othered.[14] The purity logics of white environmentalism, Ray reminds, are rooted in a eugenicist orientation that perpetuates racist and ableist systems. White environmentalism thus is not an innocuous approach that can be simply "moved to the side" or decentered, and interdisciplinarity that stays within white-dominated fields is no remedy. Across these fields and disciplinary approaches, white environmentalism needs to be taken apart, uprooted from the practices and conventions of the field. An environmental humanities oriented toward environmental justice considers the epistemic premises built into each content choice in the curriculum and each choice in praxis. This orientation also closely examines the configurations of knowledge within institutional spaces: who and what is centered, who and what is assumed to "belong," who and what is considered to be supplemental, suspect, or disposable.

Epistemic injustice and material injustice

Epistemic justice means asking what logics make environmental injustice possible; it means asking whose knowledges aren't accounted for, whose ethos is discredited, who goes unheard and unrecognized. Justice entails a recalibration of participation, recognition, procedures, distribution of resources, redress of harms, reparations, repatriation of land, and more; it also necessarily includes attention to the epistemic.[15] The objection might be raised that the matter of epistemic justice amounts to a mere thought exercise—that epistemic justice doesn't translate into the end to extractivist neoliberal economies, but this objection misses the point: epistemic justice is a matter of livability, legibility, self-determination, and decision-making power. Epistemic justice is about who counts, who matters, and on whose terms. It is the mistaken lines drawn by Cartesian dualism that prompt one to think of epistemic justice as separate from material oppression, severing mind from body, the cognitive empire (Boaventura de Sousa Santos's term) from the material one. Settler-colonial ideologies *are* their state apparatuses; the ideologies of racial capitalism *are* their industrial complexes, as they together are embroiled in cycles of reproduction. Material structure comes with epistemological structure, as mutually reinforcing. Colonial logics perpetuate the sustained invasions of Abya Yala, Tawantinsuyu, Anahuac, and many lands not named here. Colonial logics are rendered in mind-body, in relationality, and in institutional design. A recursive analysis of the epistemic and material provides a way of intervening in these logics. This analysis can enable the redress of epistemic–material violences, as it brings about a more relational understanding that prioritizes respect for ways of knowing and ways of being in their full diversity—a respect that represents a full disruption of neoliberal world orders.

All this is at stake when we chart the forms of intellectual discrimination that take place in the academy. Kuokkanen identifies how intellectual discrimination is definitive of the university, as the university has been from the start an arm of cultural and epistemic imperialism that asserted the ideology of the superiority of European knowledge.[16] To halt this continuing process will require denuding the supremacist thinking that is formative of the academic identities most privileged in the academy. Kuokkanen suggests some of what will be necessary to do this work and to discontinue deleterious institutional practice:

> First and foremost, the academy will have to acknowledge that it is founded on very limited conceptions of knowledge and the world. Because of this constrained perception, the academy cannot grasp or even hear views that are founded in other epistemic conventions.[17]

This fundamental failure of recognition plays out in daily practices: on academic senate floors, in policy decisions, in funding allocations, and in assessment practices. This is more than mere myopia; it is a willful ignorance and dismissal of knowledge practices by white academic supremacy.

Intellectual discrimination occurs in the denial of Indigenous land claims, the disparagement of cultural healing practices and medicinal understandings that are directly tied to the world's flora and that invoke spiritual realms, in the failure to recognize the lived knowledges of communities, and the belief that "people are incapable of enacting their own worlds or of realizing sociality on their terms," the latter being what Stephen Nathan Haymes identifies as characteristic of settler-plantation epistemologies that are premised on anti-Blackness.[18] A faculty member at my institution was quoted at a recent meeting stating: "Ultimately, if we have to choose between honoring someone's culture and teaching them the truth we need to choose the truth." The delegitimization of cultural knowledges in the face of a sanctioned truth arrives explicitly and implicitly in many conversations, policies, and classroom praxes across the university. The adherence to a supreme (perhaps science-guided) "truth," fantasized as untouchable by culture, only underscores the white supremacy that insists on its own universality, enforcing assimilation or, for those who do not or cannot comply, rendering disposable. White ideologies are continually reinforced as "legitimate, objective, and socially acceptable,"[19] while counterdiscourses are met with denial, sometimes through the rhetorical co-optation of the terms by which counterdiscourses are mobilized: take, for example, the ironic framing of social justice as "policing" or "canceling" by those who see social justice as an affront and who police and jeopardize the instigators of that change. These terms are trotted out as a way of ending the conversation, by those who have the power to do this silencing.[20]

Asserted neutrality and "universalism," which white supremacist thinking dwells within, is a core premise of many disciplines—not only within the STEM fields but also in other disciplines that rest in a canon of thought. This canon of thought is protected and prioritized over students' critique or refusal of it. The demand is that students accommodate disciplinary thinking. When students "fail" to perform disciplinary learning outcomes, it is the students who are imagined to be in error, not the outcomes. This is, in part, what bell hooks names as the "onslaught of biased dominator thinking" in educational institutions.[21]

To counter dominator culture involves every aspect of what we do; it is not merely a matter of content—what we teach, the ideas we proffer. As Mohanty notes,

> Racism, sexism, and homophobia are very real, day-to-day practices in which we all engage. They are not reducible to mere curricular or policy decisions—that is, to management practices. In this context we need to actively rethink the purpose of liberal education in antiracist, anticapitalist, feminist ways. . . . Thus cultures of dissent must work to create pedagogies of dissent rather than pedagogies of accommodation. And this is a task open to all—to people of color as well as progressive white people in the academy.[22]

How can we carry the call to honor the radical heterogeneity of the world's epistemologies, to counter the "structural domination and suppression—often violent—of the heterogeneity of the subject(s) in question"?[23] This question addresses not only what happens in the mind; this question is not just about interiority. Structures of belief are material structures. Belief builds; belief acts. Hegemonies are asserted and reinforced.

There is no material change without epistemological change. Moreover, the immaterial—i.e., cosmological, epistemological, and cultural—realms have been sites in which colonialism has been enacted. As Helen Callaway notes, in a passage quoted by Mohanty: "imperial culture exercised its power" through "its comprehensive symbolic order which situated permissible thinking and action and prevented other worlds from emerging."[24] Echoing the point, Linda Tuhiwai Smith explicates how the denigration of knowledge and culture "are as much a part of imperialism as materials and military strength."[25] And even militarism too has an epistemological dimension: as Melamed notes, global resource wars are also paradigm wars.[26] Imperialism continues, through racial capitalism and epistemic violence, to enact material injustices.

The work of eradicating racism and redressing continued colonialism should be central to what we teach, inseparable from any discussion of environmental degradation. Speciesist thinking is related to white supremacist, racist, colonial thinking. In the environmental humanities; our work is to counter them. As Tiffany Lethabo King notes, our work is to "refuse the necropolitical epistemological systems." These structure not only authoritarian regimes but also, in King's view, "white liberal humanist ways of thinking and imagining the world."[27] This transformative refusal requires challenging the very foundation of white environmentalism, which has rested, in Stephen Nathan Haymes' words, upon an "Eurocentric globalism and colonizing ethos that proclaim that there is only one earth ethics, and it is white."[28]

The limits of inclusive representation

To counter eurocentric globalism requires a fundamental recognition of difference. However, recognition is not sufficient. To recognize the epistemic difference and affirm it within a framework of reparations and decolonization entails a challenge to the current world order.

Cultural traditions do not exist easily side by side, and any attempt to present them in these terms risks normative pluralism, as argued in the previous chapter. It is not possible to innocently bring together multiple funds of knowledge into an easy praxis within the humanities. That is, in Cecilia Rios-Aguilar and Judy Marquez Kiyama, "funds of knowledge do not occur outside of systems of oppression; they are constructed inside an ongoing struggle for and against domination."[29] Any pluralistic orientation must be oriented toward this struggle. The imagination and enactment of new systems that condition living is paramount.

The recognition of difference alone is not sufficient, but it also is an essential, if incomplete, part of revolutionary change. With this comes the recognition of diverse environmentalisms, the situated ways in which humans are understood in relation to their environments. I take Anna Gade's suggestion that a comparative study of environmental*isms*[30] is necessary—a comparative approach that insists on an analysis of power.[31] She writes, "the relative invisibility of Muslim environmentalism in environmental humanities is the result of limitations that have a long history in European imaginaries of an Islamic other and . . . enduring presuppositions about what counts as environment."[32] Gade calls us to insist on questioning how the humanities are qualified and legitimized by relying upon and adhering to eurocentrism.[33] In other words, on what epistemological grounds are the humanities, as practiced, built? What histories does the humanities invoke and, in turn, reinforce?

The environmentalisms of Black, Indigenous, Latinx, and many peoples have been subordinated, demeaned, or unrecognized. Take, for example, Stephen Kellert's book *The Value of Life: Biological Diversity and Human Society*, which portrays African Americans as the only group that does not value "living diversity.' Stephen Nathan Haymes elucidates the racism underlying Kellert's study.

Haymes furthermore critiques the legacy of comparative analysis (as a method that Kellert employs), rightfully demonstrating how the terms of analysis, and terms of recognition, are set to reproduce white supremacy. Comparative models tend to rely on a representational pluralism that often does little to undo the conditions by which worldsenses must jockey for legitimacy.

These models are also typically conditioned by one-world thinking. AnaLouise Keating writes of one-world thinking or mono-thinking:

> When our politics are driven by mono-thinking, we assume that there is an absolute "best practice" and "best theory" to effect change. This assumption compels us to focus too narrowly on this single approach . . . Rather than work with those who use other approaches and draw on different theories, we judge them as somehow inferior—as lacking. (*Too often, of course, this judgment seems logical because our potential colleagues have defined their own approach as the best.*).[34]

Keating demonstrates the importance of a capacious approach to the central exigencies of our time. Keating finds that forms of discrimination and exclusion can exist in any model, and the work of putting these tendencies into check requires dismantling dominant epistemologies (or what Keating calls "mono-thinking" that insists on a singular and patronizing worldview). This points to what is at stake in the politics of representation and recognition.

On the question of representation Glen Sean Coulthard is an essential thinker. Countering the "liberalized appropriation of Hegel that continues to inform many contemporary proponents of identity politics," in the book *Red Skin, White Masks* Coulthard draws toward Fanonian apprehensions of recognition: showing that recognition is not a source of freedom and dignity for the colonized, but rather should be understood "*as the field of power through which colonial relations are produced and maintained*."[35] Terms of recognition are set to keep power relations in place. "[W]hen delegated exchanges of recognition occur in real world contexts of domination," Coulthard writes, "the terms of accommodation usually end up being determined by and in the interests of the hegemonic partner in the relationship."[36] Coulthard elucidates this reality in terms of the Canadian Truth and Reconciliation Commission which serves to sequester the abuses of settler colonization to a long-gone past, rather than addressing the continued violence of the settler-colonial state. Reconciliation is positioned as a means of "*overcoming* the harmful 'legacy' left in the wake of past abuse while leaving the *present* colonial rule largely unscathed."[37] Reconciliation occurs at the level of the discursive and affective, enacting a politics of *recognition* that is ultimately in service of the assuaging of settler guilt. A politics of recognition in the academy can readily function along parallel lines, as it is mobilized in scholarly and pedagogical practices when certain forms of knowledge are "let into" a dominant paradigm, recognized only to be read within dominant frameworks and in service of entrenched power relations.

Theories of the pluriversal attempt to move beyond the feeble and insidious operations of recognition. In a pluriversal approach, dominant paradigms would come undone in the presence of counterdiscourses. This fundamental unsettling would be marked and unmistakable. Thus, it would far exceed mere "inclusion" in a syllabus or course design. Here, we see the limits of representation as a model: simply "including" different orientations is insufficient, given its modes of containment.[38] Inclusion models tend to essentialize radical differences and fail to redress power disparities and continued harms. Moreover, as Mignolo makes clear, incorporation is not what is sought by marginalized communities:

> When those who have been the target of colonial and imperial subjugation and made "others" and "barbarians" assert themselves in fullness, their claim is not to be integrated into the Western proclaimed "humanity of the same" but to delink and assert "humanity in difference".[39]

To insist on inclusion into the system of "the same" is to do profound violence to those who that system was built to denigrate. The claim is instead to assert self-determination and autonomy and to halt the perpetuation of systems of violence.

The limits of empathy-based models

Limited recognition models dovetail with empathy-based models in humanities education, which assume that inequity can be resolved by a reorientation of the heart. These models prioritize the feeling, and ultimate absolution, of beneficiaries of dominant systems. The white person demands to know in order to "feel with," to empathize, which is a proxy for the disavowal of complicity. As Caitlin Breedlove, a trainer at the Highland Center activist training space, notes in an interview with Jordan Flaherty, "The well-intentioned are always reaching out. 'What is your culture? Tell me about your inner pain.' It's forced intimacy. It's not wanting to turn the reporting back on yourself. How can you ask yourself a different set of questions?"[40]—questions such as: "What are the actions for social justice and movement building that don't centre you [the well-intentioned white person] as a protagonist?" How often do white formulations of research and teaching fall into this "tell me about your inner pain" orientation? This trope represents an example of intellectual disparity in that it keeps in place a hierarchical relationship between the one who demands to know and the one who provides the raw experience. The "knower" is ultimately invulnerable before the "known" no matter how brokenhearted that knower may be in empathy.[41]

The demand to know and white presumptions of "fellow-feeling" reproduce objectifying structures. Epistemic justice requires countering these tendencies. Sara Ahmed notes how "Justice might . . . preserve the right of others not to enter into relationships, 'not to be with me,' in the first place. The other, for example, might not want my grief, let alone my sympathy, or love."[42] Not only must freedom from the imposition of fellow-feeling be insisted upon within a justice framework, but Ahmed also argues that "fellow feeling" may be not only fraught but also circumstantially impossible—at least in the case of pain; "we cannot feel another's pain as they feel it."[43] Pain cannot be shared, and much less resolved, through empathy. Ahmed calls for a response to a pain that prioritizes attentive hearing, action, and collective politics based on "learning to live with the impossibility of reconciliation, or learning that we live with and beside each other" with the knowledge that "and yet we are not as one."[44] In the environmental humanities, our work is to teach this "learning to live with"—a form of coexistence that is vigilant in redressing continued colonialist-capitalist violence without resorting to a demand for fellowship.

Colonialist assumptions

An epistemic justice orientation to the environmental humanities goes beyond common discussions of epistemology in research methods classes (which might include a survey and discussion of positivism, post-positivism, etc.). Epistemic justice is a framework for rethinking all that happens under the rubric of the environmental humanities and across interdisciplinary activity—transforming educational systems so that they cease reproducing disparities and instead create spaces where all students can thrive, fulfilling the promise of the environmental humanities by dislodging dominant ways of knowing that perpetuate social-environmental harms. Melissa K. Nelson explains central precepts practiced by multiple Indigenous nations, in naming the beliefs that:

> for humans to get along with each other and to respect our relations . . ., we must embrace and practice cognitive and cultural pluralism (value diverse ways of thinking and being). We need to not only *tolerate* difference but . . . [understand] cultural diversity as an essential part of engendering peace.[45]

Consider, for example, epistemologies regarding sustainability. Discussions of sustainability are regularly trapped within disciplinary frames (falling into the phenomenon: *if all I have is a hammer then everything is a nail*—what we know to do determines what we think should be done). A discipline-based model often negates the reality that "sustainability is premised on social and ecological bonds"[46] and interdependence across radical differences. When discussions of sustainability are premised on an "attempt to 'fix' an impenetrable and unpredictable nature,"[47] they reaffirm supremacist orientations of control. In contrast, if all flora and fauna are understood to exist in their own wisdom, then they are understood as animate life forces that are worthy of respect in an interdependent relation; we find an epistemology for sustainability that may be not only more culturally *relevant* for some (in the acknowledgment of animacy)[48] but also may do the work of disrupting holdover paradigms that are outmoded in their failure to accurately represent the complexity and unpredictability of the living world.

There are assumptions embedded in the very terms we use. The idea of a "worldview" itself represents a particular epistemic configuration. Kuokkanen cites Walter J. Ong's finding that the term is a product of technologized societies: The world of oral societies "is not so markedly something spread out before the eyes as a 'view,'" Ong writes, "rather, it is experienced and understood through a combination of senses."[49] "Worldview" as a signifier privileges visual senses over others. "Worldsense" carries different meanings, a term Mignolo prefers over "worldview"—the latter being, in his view, that which restricts and blocks the "affects and the realms of the senses beyond the eyes."[50]

"Worldview" may be aligned with a colonial understanding that "imagine[s] the world as an inanimate surface to be occupied"; this is in contrast to ideas of place-as-living-inhabitance, or place as what actively and dynamically co-creates a worldsense, with worldsenses being inextricable from the places from which they emerge and with which they exist in a living relation.[51] Worldsense, understood as being tied to the specific worldedness of each place, allows for resistance of capitalist-colonialist ideas of interchangeability and disposability. If a worldsense only exists as a particular emplacement or "being of" land, it fundamentally cannot be moved or appropriated into other contexts.

Basic colonial assumptions play out in environmental discourses, in, for example, the assumptions exemplified by Paul Wapner in the claim: "Most of us tend to think about the environment as the nonhuman world that exists 'out there'."[52] This separation of human from nonhuman is core to epistemic presumptions about wilderness, the ideologies of terra nullius that were manifest in colonialist incursion and genocide, ideologies that saw Indigenous peoples as less-than-human and mere collateral on land that too was objectified and stripped of dignity and life.

This dehumanization is fundamental to the establishment of western modernity. Speaking primarily of the African diaspora, Sylvia Wynter also presents an account of the colonial process. She demonstrates how a dominant ethnoclass (i.e., western bourgeois), by overrepresenting itself as "Man," stratified the world into a human/subhuman distinction, with Man as the only legitimate political subject. In turn, this dominant class has enacted the *epistemological disregard* for those dehumanized subjects as part of "the sustainability of the new mode of being human." Subsequently, learning, whether at the microlevel of the individual or the macrolevel of society comes to function within this regime of truth, Wynter writes. This is the continued "Coloniality of Being/Power/Truth/Freedom" that invented a genre of "Man as the Rational Self" within reigning orders of what counts as rationality.[53]

Honoring multiple ways of knowing

What would it mean to counter the tendencies of colonial thought, to find learning outside of the fundamentally dehumanizing structures of western knowledge? Students seek teachers of their communities who know the means of knowledge-seeking that resist colonial paradigms. As Robin Wall Kimmerer notes in her essay "Mishkos Kenomagwen, *the Lessons of Grass*," "Learning from experienced community members is . . . a fundamental component of Indigenous pedagogy."[54] She names some of what is taught, as a form of knowledge revitalization: careful observation, direct experience, fasting, language revitalization, restoration of ceremonies, and the protection of intact ecosystems.[55] These skills arise from an understanding that knowledge is always coupled with responsibility and the "recognition that knowledge is not our own but is the collective wisdom of the living world to which we are accountable."[56]

Kimmerer is a scientist. Too often processes such as what Kimmerer describes are relegated as "other to" science, as traditional ecological knowledge that is apart from, and suspect in relation to, science. To counter this hierarchization, Gregory Cajete calls for an expansion of what is understood as science, "to make it the whole and comprehensive form of human knowledge it needs to become,"[57] while also remaining attentive to the workings of power across this mapping of the whole.

Coming-to-terms with the operations of science-as-category entails an ongoing analysis of the epistemologies that are operating in environmentalist proposals. This means recognizing what the analytics of western science foreclose and what they have limited capacity to know. As Shawn Wilson notes in *Research Is Ceremony:*

> analysis from a western perspective breaks everything down to look at it. So you are breaking it down into its smallest pieces and then looking at those

> small pieces. And if we are saying that an Indigenous methodology includes all of these relationships, if you are breaking things down into their smallest pieces, you are destroying all of the relationships around it. So an Indigenous style of analysis has to look at all these relations as a whole instead of breaking it down, because it just won't work.[58]

The atomizing epistemology of western science described by Wilson clearly is not built to comprehend all things. It may be useful for some things. Kimmerer argues that what she calls the "language" of western science can be a useful complement to Indigenous science and ways of knowing. She argues that western science can polish our seeing—that we can "see with" multiple epistemologies, including the modalities available in western science.[59]

But it is only when the disciplines of western science are understood in the context of many other languages, understood to be built for some acknowledgments and fundamentally inept in the face of others, that it can be useful. When it is maintained as superior and all-encompassing, it performs a fundamental epistemic violence that is directly tied to cultural genocide and ecological destruction. It produces the violence of forgetting,[60] as Vanessa Watts notes, writing to fellow Indigenous peoples:

> Even amongst ourselves it can be easy to forget that our ability to speak to the land is not just an echo of a mythic tale or part of a moral code, but a reality. Whether this forgetting has been forced upon us, or our ears have become dull to the sounds of the land speaking up through our feet, it is now incumbent upon us to remember. This is not a question of "going backwards," for this implies there is a static place to return to . . . it is not a question of accessing something, which has already come and gone, but simply to listen. To act.[61]

Watts demonstrates disobedience to the analytical imposition of universalized structures of knowing. This is what Walter Mignolo calls an "epistemic disobedience" that delinks "from the illusion of the zero point epistemology."[62] In other words, it means calling into question the epistemology that transcends all others as the fundamental "objective" and universal ground, the zero point by which all other knowledges are measured.

Multiple forms of knowledge-making must be legible: knowing through story, through the planting and preparing food, through the gathering and grinding of herbs, through the intergenerational distributed consciousness of a dance circle, and through the rituals of embodied remembrance. There are layers of survivance[63] in these ways of knowing. The matter concerns not only what environmentalist discourses prioritize and pay attention to, but also ontology—what these discourses fundamentally count as existing: the medicinal properties of a plant, the lifeforce of a river, and the animacy of the sweetgrass.

As Escobar notes, environmental conflicts always have these ontological questions embedded within them. His work examines how conflicts "arise from the unequal encounter between worlds, as in so many conflicts involving resource exploitation today."[64] Resource-based conflicts involve a constellation of onto-epistemic questions about what and who counts, what and who matters, in terms of ways of being, sacredness, subsistence, cultural persistence, safety, and so on.

Epistemic justice in policy-making

Environmental exigencies are onto-epistemic dilemmas. The globe's most pressing exigencies all have within them ontological questions. The frameworks within which these exigencies are debated are too often set by those who have the power to determine the terms. The questions asked determine the answers given.

Exemplifying this point, Shalanda Baker's book *Revolutionary Power* examines the inequities of energy systems as they are produced by the disavowal of human experiences. She notes, "the energy system, again and again, works against poor people, Indigenous people, Brown and Black people" due to the assumptions that are built into these systems.[65] Issues of justice become backgrounded in white urgency narratives that question "whether entering equity in the fight against climate change is realistic given the short timeline that remains to act on climate."[66] These narratives affirm "a discourse that centers the primacy of *action* over the primacy of *people*" which, as Baker notes, "is inherently unsustainable and will inevitably lead to even greater instability than climate change portends."[67] We see this white urgency orientation play out in multiple arenas and not only in energy conversations. When climate change is approached without a justice framework, it falls into the same failures that Baker elucidates here.

As LeiLani Nishime and Kim D. Hester Williams make clear: "Despite assertions that environmental disaster is a universal threat, its effects are felt most profoundly by poor people and people of color." The dominant, racialized views of these groups as docile, unsophisticated, uncivilized, and unappreciative of the natural world serve to justify their continued endangerment in validating the idea that marginalized communities have created, or are at least complicit in, the degradation of the environment. These racist assumptions play out in the environmental humanities every time those most affected by environmental harms are not guiding the conversation in ways that are rewarded and compensated by institutional structures. As Nishime and Williams note, "dispossessed communities preserve memories, share knowledges, and enact ways of being that reflect an awareness of, and resistance to, the detrimental consequences of enduring structures of colonialism, imperialism, and neoliberal capital accumulation."[68] Environmental knowledge is not an abstraction; it is lived. Lived knowledges are not available for the taking, but they are the foundation of just transition. Those of us who are part of the environmental humanities have signed on to do the work of bringing our globe into a just and sustainable future. There are many knowledge holders, cultural practitioners, and intellectuals who can lead this work forward.

Dina Gilio-Whitaker's book *As Long as the Grass Grows* provides additional examples of onto-epistemic conflicts and barriers in environmental decision-making:

the repetitions that continue cultural-environmental deprivation or the forced removal and destruction of the "conditions necessary for life and the continuance of cultural existence"; the failure to protect sacred sites as an environmental justice issue; misunderstanding of Indigenous practices, including subsistence hunting; the "triple evils" of cultural appropriation, genetic modification, and patent ownership that rob Indigenous peoples of cultural, intellectual, and physical sustenance and ripple out to broader harms that manifest in disease and difficulty; the persistence of conservation models that perpetuate erasure of Indigenous peoples; the refusal to recognize coloniality and colonialism as still governing the relationship between the United States and Indigenous peoples (not a "wrongdoing" or "mistreatment" of the past); complacency with legal systems that represent a "rigged game against the environment and Indigenous communities"; the failure to acknowledge the connections between ecocide and genocide.[69] Colonially based racism continues every time Indigenous peoples are excluded from climate change talks and other decision-making spaces. As Gilio-Whitaker notes, "Indigenous and fourth world people [are] on the frontlines of climate change," experiencing "loss of land due to sea level rise, desertification, drought, disruptions to subsistence-based food systems, intensifying storms, loss of sea ice, and a host of related ecosystem changes"; yet Indigenous peoples are excluded from decision-making processes in response to these circumstances. The fateful effects of this exclusion are many: for example, the Kyoto Protocol's creation of a market-based system of carbon trading, Gilio-Whitaker notes, exposed Indigenous peoples to new abuses by States.[70]

The environmental humanities can make such enacted assumptions available for scrutiny and change. By centering the action and thought of "those who are on the front lines of climate change—island nations, low-income communities, communities of color, and Indigenous communities," we can find each of our roles in climate justice movements, learning what justice actually requires.[71] It matters whose vantage point comes first in the discussion—who and how we follow. When BIPOC communities and communities of the Global South host the discussion, then the epistemes represented by these groups can no longer be treated as "exceptions, deviations, and deficiencies." When they lead and frame the discussion, these epistemes are not "perceived as mere supplements . . . and as relevant only to the extent that they have something to offer to existing theories and discourses," as they typically are, Kuokkanen notes, in predominantly white spaces such as the university.[72] These epistemes do not fit within the dominant, and therefore, they cannot be supplemental to it. They do not support, and therefore, they cannot be additional to the dominant. True respect for incommensurability puts under scrutiny the terms we use to engage.

Take, for example, western discourses regarding human rights. Human rights orientations arguably impose specific understandings of individualism that are incommensurate with a relational epistemology that prioritizes interrelatedness. As Kuokkanen explains: If self cannot be understood apart from specific land relations, then a human rights framework that conceptualizes human life as distinct and isolated exists in an epistemic framework that is incompatible with many Indigenous understandings of self. The idea of individual rights imposes specific understandings of what legitimizes self-in-world that may negate Indigenous ways of knowing. As Kuokkanen notes, "those economic ideologies which focus on individual rights, freedom, and choice . . . conflict with those notions of collective solidarity that are so fundamental to Indigenous value systems."[73]

The codification of the rights of nature may go some lengths to ameliorate these limitations, but even when rights orientations embrace the more-than-human, the concern remains that rights orientations reify state-power and "hypostasize liberal-capital democratic states (as guardians of such rights)."[74] Rights discourse readily folds into a neoliberal frame: the "functioning and legitimacy of certain rationalizing modes (e.g., corporate individualism), construals of value (e.g., the sanctity of private property rights), and expression of personhood (e.g., people as consumers) that many Indigenous people do not share."[75] Melamed makes this point with Aihwa Ong as her interlocutor. Rights discourses, while on the surface proclaiming equal standing among all, also tend to obscure what Ong identifies as "graduated sovereignty" and "differentiated citizenship," meaning "the differential treatment of populations in relation to ethnoracial differences, and the dictates of developmental programs." Ong continues, further defining differentiated citizenship: "Segments of the population are differently disciplined and given differential privileges and protections in relation to their varying participation in globalized market activities."[76] A discourse of universal human rights may conceal the ways in which these disparities need to be addressed with reparative and compensatory means. The problems of differentiated citizenship will need a differentiated response.

The point here is not to negate the value of a declaration of universal human rights; rather, our purpose is to grapple with what rights discourse may elide, knowing that rights discourse is embedded within certain epistemological assumptions. The analysis that is called for is persistent: recognizing the frames that are brought to bear in each instance and locating what is beyond them. This provides a window into the utter complexity of doing coalitional work across differences, the challenge of finding models of coexistence that do not subsume radical differences and that seek restitution and reparations for the disparities that are the legacy of colonial rule.

How this translates to the environmental humanities is a question that must remain with us, a question with many answers: The coalitional orientation we seek means countering deficit narratives and research frames that rely upon constructing "entire communities as depleted"[77]; countering the "at-risking/asterisking" of Indigenous peoples and people of color; countering essentialism; countering forms of appropriation of "the other's pain"; rectifying the stolen legacies of peoples around the globe.[78]

Troubling the survey approach to learning

The frames of knowing mobilized in the environmental humanities demand our attention, in order to recognize where our attentions fail. What lies beyond the scope of our "knowing"? What cannot fit in the frames we've built?

To pose these questions is not to call for a comprehensive curriculum or coverage model that attempts to traverse the globe and find all its epistemic parts. In the face of the radical plurality of worldsenses and forms of meaning-making, there may be the academic inclination to "survey"—but this inclination is fraught. The survey course, with its sweeping attempt to cover broad swaths of knowledge, reminds of the surveyor, mapping landscapes for his own purposes. One can readily see the colonial orientations of this thinking. Merely attempting to include or, worse, "cover" in totality, multi-dimensional ways of knowing in an existing pedagogy has the same problematic effects as tokenistic pluralism. Yet, this remains a pervasive tendency in capitalist-colonial institutions: to contain and objectify thought and thinkers, denying their transformative force; to commodify thought, making it available to be plundered; to pigeonhole thinkers into a particular category; and to ignore the profound, dynamic, and ongoing work of Indigenous peoples in the face of erasure narratives or racist ideas about "traditional" thinking being "out of time" in modernity.

The survey course is a risky form, as it can readily fantasize mastery, falling into what Aimee Carrillo Rowe and Eve Tuck call "narratives of conquest" which "are present, pervasive, and mostly invisible within the settler consciousness." These narratives, Rowe and Tuck write, do "profound cultural work in reminding settlers that they belong, that their place in the social order has been hard-won through the taming of savages, and confirming their status as the rightful inheritors" of Indigenous lands and ways-of-knowing.[79] Nearly every brick and policy of the western university has been built from these narratives—with learning outcomes that demand "mastery" in a curriculum that ignores positionality and what it means to make the performance of certain knowledges "required" even when these are an affront to students' cultural understandings. At the same time, the university has few mechanisms to check the ways in which ideas are co-opted, which Indigenous principles of rightful use[80] are ignored. As Silvia Rivera Cusicanqui notes,

> Through the game of who cites whom, hierarchies are structured, and we end up having to consume, in a regurgitated form, the very ideas regarding decolonization that we indigenous peoples and intellectuals of Bolivia, Peru, and Ecuador have produced independently.[81]

White settlers do not have a "right" to know all that members of a community know. A role white settlers can play instead, perhaps, is to resist the ways in which, in Tuck and Yang's words, Indigenous "[e]pistemological, ontological, and cosmological relationships to land are interred, indeed made pre-modern and backward. Made savage"[82]—without assuming to know these relationships and without taking them up as one's own.

Silvia Rivera Cusicanqui describes resistance to dominating knowledges and regimes as a battle that academics daily contribute to, on one side or another, through the forums we participate in. Referring to the role of intellectuals in the domination of empire, Cusicanqui notes,

> I believe that it is our collective responsibility not to contribute to the reproduction of this domination. By participating in these forums and contributing to the exchange of ideas, we could be, unwittingly, providing the enemy with ammunition. And this enemy has multiple facets, both local and global, situated both in the small corners of "tiny power" in our universities and pauperized libraries and in the heights of prestige and privilege. It is from these "palaces" (the university of the North) that, following Gayatri Chakravorty Spivak, dominant ideas emanate, and it is also there that the "think tanks" (suggesting of a war) of the imperial powers are located.[83]

Cusicanqui puts into check the easy credentialing of academics who pillage the knowledges and ideas of Indigenous thinkers or, with words like "*transmodernity* and *eco-si-mía*—entangle and paralyze . . . the indigenous and African-descended people with whom these academics believe they are in dialogue."[84] How easy it is to invent a term and circulate it at the expense of people groups that the term fails to recognize and see. The hierarchies of knowledge continue to insist that what is credible in academic spaces is that which is already legible in them. The fight for credibility and ethos within political regimes is a matter of self-determination, sovereignty, and continued existence. It is a terrible, terrorizing thing to speak and be rendered uncredible from the start.

Expertise and epistemic injustice

The determination of credibility is based on, and in service of, a particular habitus. White supremacy characterizes the habitus we teach, research, and work in—so credibility is assessed in relation to it. Constructions of expertise emerge from white supremacy culture. Epistemically unjust operations draw a circle around "expertise," allowing only those granted legitimacy within dominant systems to influence decisions, even as those decisions devastatingly affect the lives of many more.

Trevor Birkenholtz[85] interrogates constructs of expertise on a macroscale and cites Scott's *Seeing like a State: How Certain Schemes to Improve the Human Condition Have Failed*,[86] noting how states have "failed to take into account local knowledge, while at the same time having a blind faith in scientific and technological progress," due to preconceptions of expertise or which knowledges "count." The effects of this regime go beyond paternalism and organizational awareness gaps; they amount to a denial of the self-knowledge and self-determination of peoples under state power, and the university is complicit, with its leveraging of certain forms of expertise and its fundamental exclusion of those who would disrupt its mechanisms of power. It is these habits of automatic assessment that determine who will be invited to the decision-making table, who counts as a "collaborative" and valid voice.

Tema Okun's discussion of white supremacy culture[87] demonstrates how the maintenance of power operates on a continuum between the interpersonal and the systemic. The much-circulated account of "Characteristics of White Supremacy Culture" from dismantlingracism.org demonstrates how a line can be drawn between an individual's defensiveness and an organizational structure that is set up to protect power relations as they exist. Defensiveness, often relegated to psychologizing frameworks that individualize affective orientations, is understood in Okun's account as a systemic issue: it is a form of conditioning that maintains the conditions of its existence. White supremacy culture produces defensiveness, and then the defensiveness shores up the culture, recursively. In Okun's analysis, defensiveness is an outgrowth of what Okun calls a "one right way" culture—an extension, perhaps, of what theorists in pluriversal studies call a one-world world. Demonstrating that this culture is an inheritance of colonialist, missionary orientations, Okun defines this culture as predicated on "the belief there is one right way to do things and once people are introduced to the right way, they will see the light and adopt it" and "when they do not adapt or change, then something is wrong with them (the other, those not changing), not with us (those who know the right way)." Thus, the status quo is reaffirmed through interpersonal interactions and individual self-regard, concurrent with, and mutually reinforcing of, the systemic.

Toward academic disruption

Our classroom pedagogies are not immune from these operations. University conventions are defended, maintained against the evidence of their profound limitations: attrition, "gaps" in experiences of students' of color in comparison to white students, harms and bias incidents reported by students that regularly go unaddressed, student disengagement from course material, and many more examples.

In "The Discourse on Language," Foucault claims, "Every educational system is a political means of maintaining or modifying the appropriation of discourse, with the knowledge and the power it carries with it." The disciplinary formations of knowledge/power are relayed through the faculty in the classroom, as they introduce disciplinary practices to students. "Gaining entrance" to a discourse means accommodating it, assimilating to it, which is the continued demand of educational systems as they stand. These disciplining functions operate as technologies of the self, shaping what it means to identify oneself in academic spaces. As Foucault reminds, "Disciplines constitute a system of control in the production of discourse, fixing the limits through the action of an identity taking the form of a permanent reactivation of the rules."[88] Identities are enacted in relation to the *rules*, the operations of knowledge/power. To decolonize institutionalized knowledge regimes is an undoing of both the academic system and academic self, as they are mutually co-constitutive. This possibility of undoing is not readily available especially for those who most benefit from the system as it stands. As Escobar notes, borrowing language from Elizabeth Povinelli, "there is always an unwillingness of liberal/moderns to let themselves 'be undone.'"[89] This unwillingness is not about simple choice or selection but instead points to the fundamental embeddedness in structures that have taught the self how to be. The unwillingness is a form of inertia, pointing to what it means to be "at rest" or indurated in a system. Those who disrupt this system push up against its hardened architecture. They are told, again in Povinelli's words: "be other so that we will not ossify, but be in such a way that we are not undone, that is make yourself doable for us."[90] The dominant liberal order enjoys the refreshing new radical idea as long as that idea can be in service of the system as it stands, as long as relations of power and privilege are not disturbed.

A politics of refusal is necessary—refusing the "doable" within dominant systems, refusing to continue to funnel the burdens of transformation through the same mechanisms of social repression. Moving "toward a post-Enlightenment configuration of knowledge forms should be one of the goals of academic cultural politics in a pluriversal register," as Arturo Escobar argues,[91] so that the dominant perspective might encounter its own destabilization.

What if our classrooms drew students to the world—the many-worlded world—in such a way that white ideation could no longer remain intact? What would it mean to exist in educational spaces that had permission to dwell near the storytellers, farmers, healers, community members, and practitioners who demonstrate in their very presence the fallibility of dehumanizing scripts, those voices and eco-social spaces that can engage the "practice of remembering against the grain of hegemonic history, of locating the silences and the struggle to assert knowledge that is outside the parameters of the dominant," which, Mohanty notes, "suggests a rethinking of sociality itself" (83)? What would it mean to reconfigure the social space of the environmental humanities, to reimagine the discourse community that arrives in the desire to understand naturecultures with complexity?

If we think of the environmental humanities classroom as a space of connection—connecting to each other as participants in the space, connecting with writers and thinkers whose work can be brought into the space—we might imagine new potentialities for it. The classroom becomes a space for gathering, rather than a space for relaying a predetermined program of knowledge. It is a carved-out space for finding movement toward learning and growth and creation, which will mean differently for each person in the room. In the space, we become attentive to epistemologies, to the knowledges and ways of knowing that are welcomed and those that are expected to be left at the door, the knowledges that arrived automatically with institutional sanction, the knowledges yet to emerge.

Seeking connection that does not "eviscerate what makes us distinct,"[92] in gathering we have plurality—but that does not ignore the lines of power that push toward dominant hegemonies. The work to interrupt those lines of power is deliberate. The idea of connection itself should be read in relation to the operations of neoliberalism which, as Pierre Bourdieu notes, destroy collective structures that impede pure market logic.[93]

A connection-based orientation to classroom practice may be aligned with what Laura Rendón describes of sentipensante pedagogy: having a goal to develop self-awareness and sense of purpose; focusing on learning and connecting to one's own and each other's lives; allowing the learning to arise from personal and communal reflections on life; focusing on inner experience and self-reflexivity; and focusing on learning as well as social responsibility.[94] This work means wrestling with what we know and how we know and how our knowledges inform our being.

Our learning outcomes, our course content, are epistemic; they ask students to believe something. Rather than resolving on a set of pre-given ideas, what would it mean to offer a curriculum that is radically open? What would it mean to follow the Black feminist principles identified by Amanda F. Anderson that "allow us to introduce a pedagogy that refuses determinacy and infallibility and invokes care as a method by which to interrogate the premise and effect of our assertions"? Anderson argues that Black feminist thought is not "knowledge that replaces another," but rather "upends the very basis of knowledge itself" in order to prompt the dissolution of dominant knowledge regimes.[95] We need a transformative approach that does more than just reproduce the denominations of our current system or even merely ameliorate them. We need what will de-center and destabilize and uproot these dominant material-epistemic arrangements, while at the same time affirming kinship cosmogonies that understand land as epistemic and onto-epistemologies of those whose "existence is resistance" to the dominant. This is what is meant by culturally sustaining education. But cultural sustenance also doesn't come from me as the white-settler educator. Cultural sustenance, cultural *survivance*,[96] cannot come from the white-settler; yet, it remains a potential of spaces of radical study. My role as a white-settler in the classroom is thus not to *teach* content, but instead, to make space for students in their knowledge journeys to the extent they invite me to do so.

I can facilitate connections in the classroom space in ways that don't center me, but instead, center the knowledges that the individuals in the temporary community of the classroom most seek. This is a fundamentally responsive, emergent, and relational orientation to pedagogy, and an attempt to counter the reality that, as noted by Kuokkanen,

> much of what is taught and discussed in the university has little relevance to indigenous students, whose cultural predispositions and aspirations are often linked to much broader collective/tribal considerations, such as exercising self-government, or bringing First Nations perspectives to bear in professional and policy-making areas.[97]

Kuokkanen laments the fact that "Instead of teaching issues that are pertinent to the goals, needs, and circumstances of indigenous peoples, the academy often seems to be trying to draw indigenous people away from who they are."[98] What it looks like to counter these assimilationist tendencies is not programmatic; it is not a simple formula. There is much complexity to account for here, recognizing that "In attempting to explain indigenous epistemes in language that is foreign to them, we risk violating their integrity, because they are not easily translatable into other systems, nor can they be reduced to simple categorizations."[99] Positionality must be accounted for, and the white-settler instructor is not entitled to the learning that students may most seek.

Let us rewrite the job description of the faculty teacher to orient it toward breaking open educational systems as they have been—including hiring practices, the organization and relative privileges of different areas of study and how they are situated in the curriculum, and what it means to earn "credit" toward a degree. It is about more than classroom practice, but of course, classroom practice has a role. The work requires radical responsiveness to who is in the room with us every day that we show up to teach. Our educational spaces are too full of transformative potential—and too often now experienced as sites of violence—to neglect the decolonizing work that scholars including Kuokkanen call us toward. Yet, there is no simple predetermination of what these classrooms will look like. They are made anew by the people in them, under a commitment to pluriversality that destabilizes academic practice as it has been, exposing its fundamental violences.

Transformation

True pluriversality will be transformative. Decolonization "will require a change in the order of the world."[100] As Tuck and Yang make clear, decolonization is a break with an ongoing structure; it is not a compromise or an accommodation within that structure. "Decolonization is not an 'and'"; it "has no synonym." A theory of decolonization need not be justified to settlers or responsible for settler questions in any way: "Decolonization is accountable [only] to Indigenous sovereignty and futurity."[101] Therefore, as Tuck and Yang note, I, as a settler, am not able to claim decolonizing orientations. But my role as a teacher is to do my best to create spaces for students to find who and what they're looking for, to do the work they're called to do. The decolonial then is not something to be simply added to a syllabus; it changes learning thoroughly and consequentially.

Notes

1 Chandra Talpade Mohanty, *Feminism Without Borders: Decolonizing Theory, Practicing Solidarity* (Duke UP, 2003) 184.
2 Given this history, nostalgia for a "return to the so-called 'Golden Era' of the university" is a tacit acceptance of the university system being "underwritten by militarized funding priorities, nationalist agendas, and incorporative projects of counterinsurgency" (Boggs, et al. 5). In their abolitionist manifesto on the university, Boggs et al. note how this nostalgia appears, in overt and less overt ways, in critical theory and in critical university studies conversations. The nostalgia manifests a white assertion of entitlement to a "time before" that is built on oppressive systems.
3 Kevin Quashie, *The Sovereignty of Quiet: Beyond Resistance in Black Culture* (Rutgers UP, 2012) 14.
4 angel Kyodo williams, Lama Rod Owens, and Jasmine Syedullah, *Radical Dharma: Talking Race, Love, and Liberation* (North Atlantic, 2016) 102, 119.
5 Laura I. Rendón, *Sentipensante (Sensing/Thinking) Pedagogy* (Stylus, 2008) 24.
6 Rendón, *Sentipensante*, 26.

7 Eve Tuck and K. Wayne Yang, "R-Words: Refusing Research," *Humanizing Research: Decolonizing Qualitative Inquiry with Youth and Communities* (Sage, 2014) 235.
8 Angela Davis, "The Fight for Ethnic Studies in CA," Webinar organized by Arab Resource and Organizing Center, 2 February 2021.
9 Leanne Betasamosake Simpson, *As We Have Always Done: Indigenous Freedom Through Radical Resistance* (U of Minnesota P, 2017) 23.
10 McKittrick writes: "Discipline is the act of relentless categorization. In many academic worlds categories are organizational tools; categories are often conceptualized as discrete from each other. Categories are things, places, people, species, genres, themes, and more, that are grouped together because they are ostensibly similar. Categories are classified and ranked and sometimes divided into subcategories (genus). . . . Canons and canonization are very clear and obvious examples of this." McKittrick, *Dear Science*, 35. Denise Ferreira da Silva echoes the point: "Post-Enlightenment Reason comprehends what it engulfs, which is only what separability, determinacy, and sequentiality can work with" (44).
11 Rhonda Magee, *The Inner Work of Racial Justice: Healing Ourselves and Transforming Our Communities Through Mindfulness* (Penguin, 2019) 52.
12 Jodi Melamed, *Represent and Destroy: Rationalizing Violence in the New Racial Capitalism* (U of Minnesota P, 2011) 102, 23.
13 Mohanty, *Feminism Without Borders*, 170.
14 Sarah Jaquette Ray, *The Ecological Other: Environmental Exclusion in American Culture* (U of Arizona P, 2013) 1.
15 Part of this definition comes from David N. Pellow, *What Is Critical Environmental Justice?* (Polity, 2018) 11.
16 Rauna Kuokkanen, *Reshaping the University: Responsibility, Indigenous Epistemes, and the Logic of the Gift* (U of British Columbia P, 2007) 13, 14.
17 Kuokkanen, *Reshaping the University*, 3.
18 Stephen Nathan Haymes, "An Africana Studies Critique of Environmental Ethics," *Racial Ecologies*, edited by LeiLani Nishime and Kim D. Hester (U of Washington P, 2018) 44.
19 Gina Ann Garcia, *Becoming Hispanic-Serving Institutions: Opportunities for Colleges and Universities* (Johns Hopkins UP, 2019) 18. Garcia is here writing in reference to Eduardo Bonilla-Silva and Tukufu Zuberi, *White Logic, White Methods: Racism and Methodology* (Rowman & Littlefield, 2008).
20 This has a parallel in the notes that Tema Okun has compiled across several trainings and facilitation sessions, to name the characteristics of "White Supremacy Culture": that people in power run from conflict and when they are faced with discomfort, "blame the person for raising the issue rather than to look at the issue which is actually causing the problem" and equate "the raising of difficult issues with being impolite, rude, or out of line."
21 bell hooks, *Teaching Critical Thinking: Practical Wisdom* (Routledge, 2009) 26.
22 Mohanty, *Feminism Without Borders*, 216–217.
23 Mohanty, *Feminism Without Borders*, 18.
24 Helen Callaway, *Gender, Culture, and Empire: European Women in Colonial Nigeria* (U of Illinois P, 1987). Quoted in Mohanty, *Feminism Without Borders*, 57.
25 Linda Tuhiwai Smith, *Decolonizing Methodologies*, 2nd ed. (Zed Books, 2012) 61.
26 Melamed, *Represent and Destroy*, 182.
27 Tiffany Lethabo King, "Humans Involved: Lurking in the Lines of Posthumanist Flight," *Critical Ethnic Studies* 3.1 (Spring 2017): 163.
28 Haymes, "An Africana Studies Critique," 46.
29 Cecilia Rios-Aguilar and Judy Marquez Kiyama, *Funds of Knowledge in Higher Education, Honoring Students' Cultural Experiences and Resources as Strengths* (Routledge, 2018) 18.
30 Anna Gade, *Muslim Environmentalisms* (Columbia UP, 2019) 15.
31 This is not a matter of relativism: not all views are equal. One is not entitled to hold racist views, to maintain racial ignorance, the dominator perspective.

32 Gade, *Muslim Environmentalisms*, 241.
33 Gade, *Muslim Environmentalisms*, 241.
34 AnaLouise Keating, *Transformation Now! Toward a Post-Oppositional Politics of Change* (U of Illinois P, 2013) 9.
35 Glen Sean Coulthard, *Red Skin, White Masks: Rejecting the Colonial Politics of Recognition* (U of Minnesota P, 2014) 17.
36 Coulthard, *Red Skin, White Masks*, 17.
37 Coulthard, *Red Skin, White Masks*, 22.
38 This is a critique of my previous work. See Adsit, *Toward an Inclusive Creative Writing* (Bloomsbury, 2017).
39 Walter D. Mignolo, *The Darker Side of Western Modernity: Global Futures, Decolonial Options* (Duke UP, 2011), 49.
40 Quoted in Jordan Flaherty, *No More Heroes: Grassroots Challenges to the Savior Mentality* (AK Press, 2016) 13.
41 Thwarting this structural relationship, some texts mock majoritized thinkers' assumption that they can come to know "the other" through interpretation. In the book *Proceed with Caution, When Engaged by Minority Writing in the Americas*, Doris Sommer shows how "minority" writing confronts this will-to-know by refusing to be read as universally relatable, universally understandable. In Sommers's analysis, these texts refuse to be incorporable into a reader's experience, they refuse the gaze of the white ethnographer who feels entitled to know—doing this through strategies such as circumlocutions, deferred translation, withheld information, secrets, absence of intimate details, refusal to confess, or a kind of unstinting availability that produces only boredom (the "featherbed resistance" that Zora Neale Hurston described). With such strategies, these texts "cripple [the reader's] authority by refusing to submit to it" (10). In this argument, Sommer borrows from Norma Klahn, in utilizing the term "particularist literature" as that which mocks the desire for the "universal." Norma Klahn proposes a "particularist" reading instead of a "universalist" one that would neither maintain the "other" as an essentialized exotic other, nor assimilate the "other" to self. "Particularist" literature protects the "minor" as sovereign and untouchable by the will to know of dominant paradigms.
42 Sara Ahmed, *The Cultural Politics of Emotion* (Routledge, 2004) 199.
43 Ahmed, *Cultural Politics*, 39.
44 Ahmed, *Cultural Politics*, 39.
45 Melissa K. Nelson, *Original Instructions: Indigenous Teachings for a Sustainable Future* (Bear and Company, 2008) 4.
46 Dan Shilling, "The Soul of Sustainability," *Traditional Ecological Knowledge: Learning from Indigenous Practices for Environmental Sustainability*, edited by Melissa K. Nelson and Dan Shilling (Cambridge UP, 2018) 5.
47 Shilling, "Soul of Sustainability," 10.
48 See Robin Wall Kimmerer, "Learning the Grammar of Animacy," *Braiding Sweetgrass* (Milkweed Editions, 2013).
49 Walter J. Ong, "World as View and World as Event," *American Anthropologist* 71.4 (1969): 634–647. Quoted in Kuokkanen, *Reshaping the University*, 57.
50 Walter D. Mignolo, "Geopolitics of Sensing and Knowing: On (De)Coloniality, Border Thinking, and Epistemic Disobedience," *Confero* 1.1 (2013): 136.
51 Arturo Escobar, *Pluriversal Politics: The Real and the Possible* (Duke UP, 2020) 27.
52 From Paul Wapner's foreword to Marie Eaton, Holly J. Hughes, and Jean MacGregor, *Contemplative Approaches to Sustainability in Higher Education* (Routledge, 2017) ix, emphasis mine.
53 Sylvia Wynter, "Unsettling the Coloniality of Being/Power/Truth/Freedom: Towards the Human, After Man, Its Overrepresentation—An Argument," *New Centennial Review* 3.3 (Fall 2003): 265, 268.
54 Robin Wall Kimmerer, "*Mishkos Kenomagwen,* the Lessons of Grass: Restoring Reciprocity with the Good Green Earth," *Traditional Ecological Knowledge: Learning from*

Indigenous Practices for Environmental Sustainability, edited by Melissa K. Nelson and Dan Shilling (Cambridge UP, 2018) 45.
55 Kimmerer, *Mishkos Kenomagwen*, 45.
56 Kimmerer, *Mishkos Kenomagwen*, 52.
57 Gregory Cajete, *Native Science* (Clear Light Publishers, 2000) 18.
58 Shawn Wilson, *Research Is Ceremony: Indigenous Research Methods* (Fernwood, 2008) 119.
59 Kimmerer, "Learning the Grammar of Animacy." This essay also usefully explicates the relationship between linguistic justice and epistemic justice.
60 See Henry Giroux, *The Violence of Organized Forgetting* (City Lights, 2014).
61 Vanessa Watts, "Indigenous Place-Thought and Agency Amongst Humans and Non-Humans (First Woman and Sky Woman Go on a European World Tour!)," *Decolonization: Indigeneity, Education, and Society* 2.1 (2013): 32.
62 Walter D. Mignolo, "Epistemic Disobedience, Independent Thought and Decolonial Freedom," *Theory, Culture, and Society* 26.7–8 (2009): 160.
63 This is Vizenor's word for "a native sense of presence, the motion of sovereignty and the will to resist dominance. Survivance is not just survival but also resistance, not heroic or tragic, but the tease of tradition, and my sense of survivance outwits dominance and victory" (*Fugitive Poses* 93, quoted in Eve Tuck, "Suspending Damage: A Letter to Communities," *Harvard Educational Review* 79.3 (Fall 2009): 422).
64 Arturo Escobar, "Transition Discourses," *Constructing the Pluriverse: The Geopolitics of Knowledge*, edited by Bernd Reiter (Duke UP, 2018) 75.
65 Shalanda Baker, *Revolutionary Power: An Activist's Guide to the Energy Transition* (Island P, 2021) 8.
66 Baker, *Revolutionary Power*, 15.
67 Baker, *Revolutionary Power*, 15.
68 LeiLani Nishime and Kim D. Hester Williams, *Racial Ecologies* (U of Washington P, 2018) 6.
69 Dina Gilio-Whitaker, *As Long as Grass Grows: The Indigenous Fight for Environmental Justice, from Colonization to Standing Rock* (Beacon, 2019) 12, 33, 39, 69, 79, 106, 130, 162.
70 Gilio-Whitaker, *As Long*, 109.
71 Baker, *Revolutionary Power*, 97.
72 Kuokkanen, *Reshaping the University*, 72.
73 Kuokkanen, *Reshaping the University*, 37. Dina Gilio-Whitaker explains, "In Indigenous worldviews, there is no separation between people and land, between people and other life forms, or between people and their ancient ancestors whose bones are infused in the land they inhabit and whose spirits permeate the place." Gilio-Whitaker, *As Long*, 27. Deborah Bird Rose makes a similar point: She critiques the epistemology that makes a singular human the focal point or apex of the analysis. An ecological ontology for Deborah Bird Rose means the recognition that all beings are enmeshed "within life-giving bonds that are inclusive. . . ." and that should be "pervaded with an ethic of care and responsibility." The organism cannot be understood apart from its environment. All beings are enmeshed in fundamental interrelatedness. An ecological ontology acknowledges that "non-humans are not only lively, but can be understood as persons, that is as human beings with an interest in their own life, their own way of life (conatus), and their own desires." Deborah Bird Rose, "Connectivity Thinking, Animism, and the Pursuit of Liveliness," *Educational Theory* 67.4 (2017): 496, 495.
74 Melamed, *Represent and Destroy*, 195.
75 Melamed, *Represent and Destroy*, 182.
76 Aihwa Ong, *Neoliberalism as Exception: Mutations in Citizenship and Sovereignty* (Duke UP, 2006) 88.
77 Tuck, "Suspending Damage," 409.
78 Tuck and Yang, "R-Words," 14.
79 Aimee Carrillo Rowe and Eve Tuck, "Settler Colonialism and Cultural Studies," *Cultural Studies, Critical Methodologies* 17.1 (2017): 6.

80 Kimberly Christen, "Tribal Archives, Traditional Knowledge, and Local Contexts: Why the 's' Matters," *Journal of Western Archives* 6.1, https://digitalcommons.usu.edu/westernarchives/vol6/iss1/3.
81 Silvia Rivera Cusicanqui, *Ch'ixinakax Utxiwa: On Decolonising Practices and Discourses* (Polity, 2020) 61.
82 Tuck and Yang, "Decolonization Is Not a Metaphor," 5.
83 Cusicanqui, *Ch'ixinakax Utxiwa*, 58.
84 Cusicanqui, *Ch'ixinakax Utxiwa*, 59.
85 Thanks to Deepti Chatti for pointing me to this reference and for allowing me to sit in on her lecture on this topic.
86 See James C. Scott, *Seeing Like a State: How Certain Schemes to Improve the Human Condition Have Failed* (Yale UP, 1998).
87 Tema Okun, "Characteristics of White Supremacy Culture," *dRworks*, www.dismantlingracism.org. Since the time of this book's writing, Okun has published a revision to the original list of characteristics. This is available at www.whitesupremacyculture.info.
88 Michel Foucault, "The Discourse on Language," *The Archaeology of Language* (Vintage, 2010; originally published in 1971) 224.
89 Elizabeth Povinelli, "Radical Worlds: The Anthropology of Incommensurability and Inconceivability," *Annual Review of Anthropology* 30 (2001): 319–334. Quoted in Escobar, "Transition Discourses," 83.
90 Povinelli, "Radical Worlds".
91 Escobar, *Pluriversal Politics*, xii.
92 Williams, Owens, and Syedullah, *Radical Dharma*, 22.
93 Pierre Bourdieu, "The Essence of Neoliberalism," *Le Monde Diplomatique* (December 1998), https://mondediplo.com/1998/12/08bourdieu.
94 Rendón, *Sentipensante*, 90.
95 Amanda F. Anderson, "On (the Question of) 'Knowledge Itself': Teaching Black Feminism Now," *Feminist Formations* 32.1 (Spring 2020): 242.
96 Gerald Vizenor, ed. *Survivance Narratives of Native Presence* (U of Nebraska P, 2008).
97 Kuokkanen, *Reshaping the University*, 52. Four Arrows' introduction to the edited collection *Unlearning the Language of Conquest* identifies "assumptions about life that many of the varied and various Indigenous Peoples generally share" (19), including that the natural world is ultimately more about cooperation than it is about competition; the concept of reciprocity can guide living systems toward balance; it is a pernicious myth that injustice is a necessary and inevitable by-product of human activity; human decisions are best made from the heart as well as the mind; that a Great Mysterious spirit is within all its creations; that living Indigenous cultures possess wisdom that is vital for all futures; and that it is necessary to acknowledge and advocate for the return of lands to Indigenous peoples; land is inextricably tied to Indigenous lifeways. This list is not meant to be essentialist: Four Arrows is highlighting principles where diverse Indigenous philosophies may overlap. This does not negate the radical differences between Indigenous communities. As Kuokkanen notes, "Different indigenous peoples have very different ways of articulating and identifying the world and thus practising their cosmologies. Yet many indigenous ontologies, however separate and distinct, share certain fundamental perceptions of the order of things, especially as these relate to the human relationship to and position in the world" (59).
In describing these things, it is important to acknowledge how, in Jodi Melamed's words, "Reductionism and essentialism must be guarded against. The knowledge systems of indigenous peoples differ greatly from one another and are not internally homogenous. They cannot be made completely transparent to culturally non indigenous peoples, nor can one indigenous epitome be transcoded seamless (or even adequately) into another, and recomprehending the world does not change it" (83).
98 Kuokkanen, *Reshaping the University*, 52.

99 Kuokkanen, *Reshaping the University*, 60.
100 Tuck and Yang, "Decolonization Is Not a Metaphor," 31.
101 Tuck and Yang, "Decolonization Is Not a Metaphor," 35.

Works cited

Ahmed, Sara. *The Cultural Politics of Emotion*. Routledge, 2004.

Anderson, Amanda F. "On (the Question of) 'Knowledge Itself': Teaching Black Feminism Now." *Feminist Formations* 32.1 (Spring 2020): 238–243.

Arnold, Matthew. *Culture & Anarchy: An Essay in Political and Social Criticism. 1867*. Macmillan, 1891.

Baker, Shalanda. *Revolutionary Power: An Activist's Guide to the Energy Transition*. Island Press, 2021.

Birkenholtz, Trevor. "Contesting Expertise: The Politics of Environmental Knowledge in Northern Indian Groundwater Practices." *Geoforum* 39 (January 2008): 466–482.

Boggs, Abigail, Eli Meyerhoff, Nick Mitchell, and Zach Schwartz-Weinstein. *Abolitionist University Studies: An Invitation*, 2019. https://abolition.university/invitation/

Bonilla-Silva, Eduardo, and Tukufu Zuberi. *White Logic, White Methods: Racism and Methodology*. Rowman & Littlefield, 2008.

Bourdieu, Pierre. "The Essence of Neoliberalism." *Le Monde Diplomatique*, December 1998. https://mondediplo.com/1998/12/08bourdieu

Cajete, Gregory. *Native Science*. Clear Light Publishers, 2000.

Callaway, Helen. *Gender, Culture, and Empire: European Women in Colonial Nigeria*. U of Illinois P, 1987.

Christen, Kimberly. "Tribal Archives, Traditional Knowledge, and Local Contexts: Why the 's' Matters." *Journal of Western Archives* 6.1, https://digitalcommons.usu.edu/westernarchives/vol6/iss1/3

Coulthard, Glen Sean. *Red Skin, White Masks: Rejecting the Colonial Politics of Recognition*. U of Minnesota P, 2014.

Cusicanqui, Silvia Rivera. *Ch'ixinakax Utxiwa: On Decolonising Practices and Discourses*. Polity, 2020.

da Silva, Denise Ferreira. "Reading the Dead: A Black Feminist Poethical Reading of Global Capital." *Otherwise Worlds: Against Settler Colonialism and Anti-Blackness*, edited by Tiffany Lethabo King, Jenell Navarro, and Andrea Smith. Duke UP, 2020.

Eaton, Marie, Holly J. Hughes, and Jean MacGregor. *Contemplative Approaches to Sustainability in Higher Education*. Routledge, 2017.

Escobar, Arturo. "Transition Discourses." *Constructing the Pluriverse: The Geopolitics of Knowledge*, edited by Bernd Reiter. Duke UP, 2018.

Escobar, Arturo. *Pluriversal Politics: The Real and the Possible*. Duke UP, 2020.

Flaherty, Jordan. *No More Heroes: Grassroots Challenges to the Savior Mentality*. AK Press, 2016.

Foucault, Michel. "The Discourse on Language." *The Archaeology of Language. 1971*. Vintage, 2010.

Fricker, Miranda. *Epistemic Injustice: Power and the Ethics of Knowing*. Oxford UP, 2007.

Gade, Anna. *Muslim Environmentalisms*. Columbia UP, 2019.

Garcia, Gina Ann. *Becoming Hispanic-Serving Institutions: Opportunities for Colleges and Universities*. Johns Hopkins UP, 2019.

Giroux, Henry. *The Violence of Organized Forgetting*. City Lights, 2014.

Haymes, Stephen Nathan. "An Africana Studies Critique of Environmental Ethics." *Racial Ecologies*, edited by LeiLani Nishime and Kim D. Hester. U of Washington P, 2018.

hooks, bell. *Teaching Critical Thinking: Practical Wisdom*. Routledge, 2009.
Keating, AnaLouise. *Transformation Now!: Toward a Post-Oppositional Politics of Change*. U of Illinois P, 2013.
Kellert, Stephen R. *The Value of Life: Biological Diversity and Human Society*. Island/Shearwater, 1997.
Kimmerer, Robin Wall. *Braiding Sweetgrass*. Milkweed, 2013.
Kimmerer, Robin Wall. "*Mishkos Kenomagwen*, the Lessons of Grass: Restoring Reciprocity with the Good Green Earth." *Traditional Ecological Knowledge: Learning from Indigenous Practices for Environmental Sustainability*, edited by Melissa K. Nelson and Dan Shilling. Cambridge UP, 2018.
King, Tiffany Lethabo. "Humans Involved: Lurking in the Lines of Posthumanist Flight." *Critical Ethnic Studies* 3.1 (Spring 2017): 162–185.
Kuokkanen, Rauna. *Reshaping the University: Responsibility, Indigenous Epistemes, and the Logic of the Gift*. U of British Columbia P, 2007.
Magee, Rhonda. *The Inner Work of Racial Justice: Healing Ourselves and Transforming Our Communities Through Mindfulness*. Penguin, 2019.
McKittrick, Katherine. *Dear Science and Other Stories*. Duke UP, 2020.
Melamed, Jodi. *Represent and Destroy: Rationalizing Violence in the New Racial Capitalism*. U of Minnesota P, 2011.
Mignolo, Walter D. "Epistemic Disobedience, Independent Thought and Decolonial Freedom." *Theory, Culture, and Society* 26.7–8 (2009): 159–181.
Mignolo, Walter D. *The Darker Side of Western Modernity: Global Futures, Decolonial Options*. Duke UP, 2011.
Mignolo, Walter D. "Geopolitics of Sensing and Knowing: On (De)Coloniality, Border Thinking, and Epistemic Disobedience." *Confero* 1.1 (2013): 129–150.
Mohanty, Chandra Talpade. *Feminism Without Borders: Decolonizing Theory, Practicing Solidarity*. Duke UP, 2003.
Nelson, Melissa K., ed. *Original Instructions: Indigenous Teachings for a Sustainable Future*. Bear and Company, 2008.
Nishime, LeiLani, and Kim D. Hester Williams. *Racial Ecologies*. U of Washington P, 2018.
Okun, Tema. "White Supremacy Culture." *DRWorksBook*, 2019. www.dismantlingracism.org/white-supremacy-culture.html
Ong, Aihwa. *Neoliberalism as Exception: Mutations in Citizenship and Sovereignty*. Duke UP, 2006.
Ong, Walter J. "World as View and World as Event." *American Anthropologist* 71.4 (1969): 634–647.
Povinelli, Elizabeth. "Radical Worlds: The Anthropology of Incommensurability and Inconceivability." *Annual Review of Anthropology* 30 (2001): 319–334.
Ray, Sarah Jaquette. *The Ecological Other: Environmental Exclusion in American Culture*. U of Arizona P, 2013.
Rendón, Laura I. *Sentipensante (Sensing/Thinking) Pedagogy*. Stylus, 2008.
Rose, Deborah Bird. "Connectivity Thinking, Animism, and the Pursuit of Liveliness." *Educational Theory* 67.4 (2017): 491–508.
Rowe, Aimee Carrillo, and Eve Tuck. "Settler Colonialism and Cultural Studies." *Cultural Studies, Critical Methodologies* 17.1 (2017): 3–13.
Scott, James C. *Seeing Like a State: How Certain Schemes to Improve the Human Condition Have Failed*. Yale UP, 1998.
Shilling, Dan. "The Soul of Sustainability." *Traditional Ecological Knowledge: Learning from Indigenous Practices for Environmental Sustainability*, edited by Melissa K. Nelson and Dan Shilling. Cambridge UP, 2018, pp. 3–15.

Simpson, Leanne Betasamosake. *As We Have Always Done: Indigenous Freedom through Radical Resistance*. U of Minnesota P, 2017.
Smith, Linda Tuhiwai. *Decolonizing Methodologies*. 2nd ed. Zed Books, 2012.
Tuck, Eve. "Suspending Damage: A Letter to Communities." *Harvard Educational Review* 79.3 (Fall 2009): 409–427.
Tuck, Eve, and K. Wayne Yang. "Decolonization Is Not a Metaphor." *Decolonization: Indigeneity, Education, & Society* 1.1 (2012): 1–40.
Tuck, Eve, and K. Wayne Yang. "R-Words: Refusing Research." *Humanizing Research: Decolonizing Qualitative Inquiry with Youth and Communities*. Sage, 2014, pp. 223–248.
Watts, Vanessa. "Indigenous Place-Thought and Agency Amongst Humans and Non-Humans (First Woman and Sky Woman Go on a European World Tour!)." *Decolonization: Indigeneity, Education, and Society* 2.1 (2013): 20–34.
williams, angel Kyodo, Lama Rod Owens, and Jasmine Syedullah. *Radical Dharma: Talking Race, Love, and Liberation*. North Atlantic, 2016.
Wilson, Shawn. *Research Is Ceremony: Indigenous Research Methods*. Fernwood, 2008.
Wynter, Sylvia. "Unsettling the Coloniality of Being/Power/Truth/Freedom: Towards the Human, After Man, Its Overrepresentation—An Argument." *New Centennial Review* 3.3 (Fall 2003): 257–337.

3

STAYING WITH THE PRACTICE

The resources of mindfulness

In the talk "The Times Are Urgent, So Let Us Slow Down," Bayo Akomolafe notes an "upsurge of wisdom" across multiple traditions in the recognition that "We are relational beings. We are essentially non-essential. We are relationships. We are not things in a vacuum, in a container. We are the earth touching itself, in a way." The reach of new materialism and agential realism, as Rosiek, Snyder, and Pratt note, has resonances with long-held Indigenous understandings of an animate, interdependent world, although scholars within new materialism and related fields too often fail to cite and continue to talk past Indigenous knowledge. Akomolafe notes that, as Indigenous peoples have long known, we cannot "think of the self as the industrial capitalist self that is atomized and separate from all the other selves."

For those embroiled within dominant systems, accessing a fundamentally relational self requires a reorientation at the level of ontology, a sharp break from the way of being that dominant orders of knowledge produce. Liberal individualism is a way of being that resists and denies relationality; it is a product of sanctioned ignorance of the permeability of our bodies, of the ways in which we are given over to the other.[1] To use the lexicon suggested in Karen Barad's *Meeting the Universe Halfway*, this sanctioned ignorance is a denial of our entanglements in favor of a fabricated notion of a self-contained and self-possessed existence, but, as Karen Barad reminds, "Existence is not an individual affair."[2] Writing this in a time of a pandemic makes all the more grave the willful ignorance of this fact, pronounced on mask-less faces and with each refused stay-at-home order. The politicization of the mask is not a result of an absence of knowledge; it instead becomes a manifestation of constructs of the insulated and isolated self, a self understood in atomistic terms and unyielding to the reality of our interconnectedness and porousness in relation to human and more-than-human selves.

DOI: 10.4324/9781003037439-4

While progressive thought may righteously decry the selfish ignorance of the maskless, we've seen that racial liberalism operates within the same individualistic order. The care of the liberal self exists in the same knowledge systems as what has produced the mask-rejecter, the climate-change denier, and so on. Within race liberalism and (neo)liberal multiculturalism, the performativity of the supposedly "progressive" ego (as "good" ally or otherwise) reaffirms its own exceptionalism. White supremacy constructs individualist selves that believe themselves to be white, that believe themselves to be right, and that shore up their own "goodness" within the systems that privilege them. It is likely not coincidental that books on how to *be* an antiracist outsell those that foreground a structural analysis on their covers. Robin DiAngelo's work, firmly placed among those best-sellers, is critiqued for its reliance on a white racial liberal ideal—ego-driven and inscribed within what Melamed terms official racism.[3]

In a talk titled "Between Pessimism and Optimism: White Crisis," George Yancy encourages white people to find themselves in crisis or to find a crisis of self. Because to be white is to be racist, we must die to whiteness, Yancy argues.[4] The metaphor of death is apt: whiteness can't be undone by criticality alone. An anti-racist whiteness cannot come into being through a simple set of principles learned. Accounting for whiteness needs something more—something that mere intellectualization is not equipped to enable, something that is fully embodied and lived.

What ontological practices can prompt this crisis of white selfhood? What ontological practices can prompt a radical break with the knowing self that has been inextricable with white settler fabrications? What practices may prompt a means of breaking open the self that has been delivered by dominant systems, nourishing those who have been most harmed by this delivery and enabling the beneficiaries of this system to break from it? Our practice needs to recognize that, as Resmaa Menakem argues, white supremacy is more than a cognitive belief system; "only a fraction of it exists in our cognitive brains," he notes. "For the most part, white supremacy lives in our bodies."[5] What forms of embodiment and relationality may enable a nodal self that participates in the collective overturning of systems as they stand, a fundamentally relational self that breaks open the lies of liberal individualism? To see self as a node in a system is to find (even of an approximation of) an egoless orientation that refuses the constructs of self-provided through whiteness.[6] To be a node is to forward and pass along the energy of a movement, to be in service of movement-building. What habits and orientations of the mind–body might allow for movement?

These questions prompt my turn to a cluster of practices commonly associated with mindfulness, taking seriously the traditions that predate modernity. However, I make this turn with full awareness that mindfulness practice is not in and of itself a "good." It is not a "way out."

As a practice, mindfulness is not impervious to the forces and conditions of our world. As Ronald E. Purser, Jeff Wilson, David Forbes, Yuria Celidwen, among others have noted, the uptake of cross-traditional contemplative practices in the west has reaffirmed the neoliberal self. Mindfulness has been proffered as "nothing more than basic concentration training," a "tool of self-discipline, disguised as self-help" that works within, and not against the technologies of self[7] relied upon in a racial capitalist system. When appropriated in western anglosphere discourse, mindfulness is often pitched as a kind of antidote—something that will give the assurance of any number of things: self-regulation, mental discipline, new cognitive scripts, emotional processing, physical conditioning, stress reduction, pain management, attentional awareness, self-reflexivity, nonjudgment, adaptability, empathy and compassion, and so on. In the United States, Jeff Wilson observes, mindfulness is employed "to gain control over mental illness, achieve self-fulfillment, and sell books."[8] This utilization to "gain control," Wilson notes, clearly indicates the detachment of mindfulness practices and pedagogies from their theophilosophical contexts of emergence; mindfulness is taken up and distorted by colonial logics of control and normativity, assumptions of universalism, neoliberal individualism, and western capitalist consumerism. Mindfulness has been commoditized to sell everything from Kentucky Fried chicken to coloring books.[9] Such is the nature of what David Forbes calls the "Minefulness Industrial Complex," (using "mine" here to emphasize the tie between the commoditization of mindfulness and neoliberal possessive selfhood). Ancient wisdom practices are secularized (e.g., Buddhist principles evacuated of their Buddhism),[10] medicalized (e.g., measured quantitatively with an understanding of anatomy in mechanistic terms), and shaved down to fit into epistemologies that are fundamentally antithetical to original teachings. Mindfulness practices are instrumentalized for better performance in a late capitalist system oriented toward optimization in consumerist terms. Ancient wisdom practices, tied to ritual and meant to facilitate connection and belonging within kinship epistemologies, are shoehorned into neoliberal's configuration of the individual. Capitalism takes whatever is in its path.

However, contemplative practices exceed capitalism's encroachment, rooted as they are in traditions that extend to time immemorial. Because of these legacies, there is the potential that contemplative practices can slow, and possibly even suspend, the ways in which we may be given over to dominant systems in both our interiority and daily cadence. Contemplative practices may have the potential to make us more supple and detached from the systems we are embroiled in. They may open a space for what Boaventura de Sousa Santos calls "a form of knowledge that aspires to an expanded conception of realism that includes suppressed, silenced, or marginalized realities, as well as emergent and imagined realities."[11] I understand contemplative practices to open up within the self a way of encountering "the impossibility of grasping the infinite epistemological diversity of the world," in Santos' words.[12]

Contemplative practice

In recent years, there has been a groundswell of interest in contemplative pedagogies in higher education,[13] with increasing interest from the environmental sciences, environmental humanities, and sustainability studies.[14] Scholars with a range of backgrounds and orientations—including Laura Rendón (sentipensante pedagogy), bell hooks (education as the practice of freedom), adrienne maree brown (emergent strategy), Ron Miller (holistic education), and Arthur Zajonc (meditation science in contemplative practice)—have contributed to contemplative studies, an interdisciplinary conversation that includes neuroscience, pedagogy, and many other areas.[15] Contemplative studies traverse the academic disciplines, and contemplative practices continue an array of cultural traditions. Michael Yellow Bird of the MHA Nation (Mandan, Hidatsa, and Arikara) writes of many Indigenous contemplative practices that incorporate principles and processes of mindfulness: he explains, "the awareness and concentration aspects of mindfulness paralleled mindfulness skills practised in tribal activities, such as fishing, basketmaking, prayer, storytelling, and the singing of tribal songs," which emphasize the importance of attending to what one is "doing, thinking, and feeling." For sacred tribal practices, it is "necessary to put good thoughts and feelings into them in order to achieve a good result."[16] These contemplative practices emphasize interdependence; mindfulness is not a solitary act and is instead a practice that renews deep connections between interdependent life forms and forces.

Contemplative practices arise from diverse traditions. These traditions are brought together in study because of shared foci: stillness, awareness, presence, body and breath, compassion, introspection, conscious listening—and the cultivation of mindful attention, what Patricia Owen-Smith calls "the *sine qua non* of all contemplative practices."[17] While there is some debate in the field regarding core definitions of mindfulness and meditation, one can provisionally define mindfulness as Yuk-Lin Renita Wong does: as an "internal, fluid process of moment-to-moment consciousness that cannot fully be captured by quantification."[18] It is presence with intention and without arresting attachment.

Rhonda Magee calls contemplation "a long, loving look at the real," with "loving" here being the realization of all the ways in which the world currently does not fulfill the promise of this term.[19] Contemplative practice is committed to connecting the inner and outer world,[20] believing in the intertwining of internal orientation and externalized action. Mindfulness, Magee continues, "at its simplest, is paying attention to life as it unfolds, grounded in the body and breath, and allowing awareness to settle in the mind to increase presence and consciousness of interrelatedness with others."[21] It moves beyond mere metacognitive reflection, as it opens space for nonconcept, for suspending the apparatuses of thought that have been installed in the mind through dominant epistemologies.

While mindfulness practices may find resonances in "habits of mind" approaches that have been popularized in Scholarship of Teaching and Learning (SoTL),[22] the intention is not to instrumentalize mindfulness practices in order to reach a particular set of learning outcomes, as is the tendency in "habits of mind" approaches. These habits of mind are also cultural and should be understood as contingent.[23]

Ideally, contemplative pedagogies would slow us down enough to truly be present to humans and more-than-humans and their intergenerational traditions, in their full multidimensionality and radical difference. Ideally too, contemplative pedagogies would engage the mind–body, recognizing that learning, thinking, and creating are embodied processes. However, too often this encounter is short-circuited to prioritize pre-determined "content" in a curriculum understood in terms of "already resolved" knowledge. Content conceptualized as such is not responsive, not emergent, and is therefore poorly equipped to prompt what has not been prompted before. The "already-resolved" has little capability therefore. Our postures, conversations, forums, and processes must be equipped to stay with the emergent, as they also stay with the traditions and heritages that have been sustaining, in order to counter any disciplining toward the "what is." When I look toward contemplative practice, I look to it for these reasons.

Contemplative practice as pluriversal

Owen-Smith argues that contemplative practice itself is pluriversal, writing that: "contemplative practices are human practices existing in every tradition throughout time. They are the core of human beings and, therefore, have an innate integrity."[24] In historicizing contemplative practices in this way, Owen-Smith does not wish to homogenize them. The tendency to detach mindfulness practices from their cultural contexts and vacate them of cosmologies and meanings is a violence to Indigenous and other epistemologies which, in Shawn Wilson's words, are "all about developing through the formation of relationships. An idea cannot be taken out of this relational context and still maintain its shape."[25]

Contemplative practices have been adopted by those who are the beneficiaries of a colonial history and fitted into colonial logics: for example, what Linda Tuhiwai Smith identifies as the tendency to "collect the intangibles: the belief systems and ideas about healing, about the universe, about relationships and ways of organizing, and the practices and rituals which go alongside such beliefs . . ." may be descriptive of contemplative practices in US higher education, an example of spiritual colonialism.[26] Removing belief systems, practices, and rituals from their contexts is an extension of the extractivist logics of continued colonial relations. This "cherry picking"[27] snatches a practice from its rootedness in culture, history, and tradition—plundering an idea, like a cherry fruit, without understanding the tree, the broader epistemology, and cosmology that sustains it. The act of "cherry picking"—taking up certain ideas by divorcing them from their rootedness in context—denies the broader linguistic, epistemological, relational, and material contexts that give an idea meaning and that create responsible relationality around it. While of course we cannot possibly represent any epistemology in totality, guarding against decontextualizing approaches is necessary.

Jeff Wilson's book *Mindful America* asks, "How did we get to this point, where mindfulness apparently can exist entirely outside of Buddhism, and non-Buddhists can teach other non-Buddhists about Buddhist-derived mindfulness . . .?"[28] The history of contemplative pedagogy can be captured with only a wide lens, inclusive of many traditions, findings, principles, and values (not only Buddhism), each configured differently in different cultures, communities, and situations. Yet too often this branching history is ignored. Practices are plucked from across multiple longstanding traditions and instrumentalized to therapize certain symptoms, without a critique of the systems that sustain them—a point made by multiple scholars and practitioners including Ronald E. Purser, who popularized the critique of McMindfulness in a *Huffington Post* article (2013) and subsequent book on the topic (2019), as well as Yuk-Lin Renita Wong who observes,

> Mindfulness has been commoditized as a cost-saving technique to be added to the Western toolbox to treat modern ailments such as stress and mental distress. This has been done without examining the socioeconomic and sociopolitical conditions that contribute to the production of stress and mental health issues in this global, neoliberal capitalist era.[29]

In such circumstances, mindfulness approaches are tailored to fit into racial capitalism and the gross inequities and violences that it perpetuates. Hence mindfulness may be used in police training, but it doesn't prevent police brutality.

Even in spaces that at least attempt to resist the will-to-commodify (Buddhist sanghas, for example), white supremacy still may take hold, as noted by angel Kyodo williams, Jasmine Syedullah, and Lama Rod Owens in *Radical Dharma* and by Zenju Earthlyn Manuel in *The Way of Tenderness*. As Manuel notes, "Identity should not be dismissed in our efforts toward spiritual awakening,"[30] yet white-dominated spiritual spaces again and again carry out colorblind orientations that refuse to grapple with race and racism.

In turning toward contemplative practice, I claim no cure-all. However, we need ways of doing-thinking-being that thwart the rote and rotten lines of thought that have been instilled by dominant epistemologies. The practices and traditions that call one toward attentive embodiment, interrelatedness, and an expansive sense of the ancestral world are necessary to the intentions described in this book.

In an effort to counter the colonial tendency toward erasure and removal, contemplative practitioners can commit to actively respecting the radical difference of coexisting languages and cosmologies that come together under the rubric of contemplative practice and seek also to trace the roots of tradition. In this, the goal is not to locate a "pure and authentic first"; at the same time, we should have a critical lens on the uptake of mindfulness as it is brought into existing ideologies. This means interrupting any easy overlay of dominant epistemologies upon contemplative practices.

Mindfulness practices are embedded in a range of epistemologies that have morphed and evolved since ancient times, arising from multiple cultural traditions: from Kemetic yoga or Qigong or Buddhism.[31] As Owen-Smith notes, "contemplative practices are definitive components in Buddhist and Taoist meditation, Hindu yoga forms, Christian contemplative prayer, Sufi metaphysical reflection and the Jewish mystical school of thought, Kabbalah."[32] Kevin Quashie, in *The Sovereignty of Quiet*, argues that contemplative practice is part of Black cultural practices exceeding religious traditions. Tracing contemplative orientations across Black literatures, he writes, "there is a strong contemplative tradition in black culture, a tradition inspired by the existential struggle of living with the confines of racial identity."[33]

These few examples of contemplative traditions are meant to illustrate radical diversity. The list is of course far from exhaustive, and these traditions are not meant to be presented as if they could be found, hermetically sealed and whole[34]; instead, these examples are meant to signal the need for honoring the ever-morphing epistemological diversity of the world. Contemplative practices are themselves pluralistic. We come to them in recognition of different practices with different purposes and inflections, as is the radical contingency of cultural embeddedness. When I introduce the Tree of Contemplative Practice (CMind) in the classroom, I contextualize it by stating that the trees of each culturally specific contemplative practice look different, as an oak tree is different from the Ghaf tree—yet there is commonality too. In this conversation, I also seek to give a sense of my own history with contemplative practices and where my practice is rooted, as I invite students to find their own history. Contemplative practices are emergent and unfolding; they have histories as they are situated differently.

Each iteration and instance of "mind" in mindfulness, "contemplation," "attention," "detachment," is nestled in an epistemology. We cannot neglect a particularized regard for each term. To do otherwise would be to further entrench universalist notions that are relentlessly homogenizing or domineering. Ghanaian philosopher Kwasi Wiredu writes of the untranslatability of the Cartesian precept *cogito, ergo sum* in Akan. "'Thinking,' in Akan, means 'measuring something,' which does not make sense coupled with the idea of being."[35] Individualistic understandings of self are incommensurate with communitarian briefs that there is no self without the other, that we are fundamentally relational. Albert Yava notes that

> if the Hopi were asked who (i.e., which individual) was to be credited for something, they often made up a name. No record of individual achievements or failures was kept, and the Hopi are a whole and not the sum of their parts.

Yava continues, "The ontological stance of belonging and being connected to others and to the world" is found in many cultural expressions. We can find this ontological stance in *all my relations* orientations among Indigenous groups on Turtle Island; in the term "ubuntu," the Nguni Bantu term for humanity; and the related Zulu phrase uMuntu ngumuntu ngabantu (roughly translated "a person is a person because of the people"). The deliberate respect for these languages, and the epistemologies these languages carry with them, remains significant.

To recognize the cultural embeddedness of contemplative practice is also to fully acknowledge accompanying spiritual legacies—doing this despite the academic tendency acknowledged by M. Jacqui Alexander in *Pedagogies of Crossing:* "There is a tacit understanding that no self-respecting postmodernist would want to align herself (at least in public) with a category such as the spiritual, which appears so fixed, so unchanging, so redolent of tradition."[36] This aversion to, and exclusion of, the spiritual should give us pause, especially as we understand western academia as an arm of colonialism: its classrooms used as a tool of assimilation, stratification, and domination[37] and its version of research "inextricably linked to European imperialism and colonialism" as Linda Tuhiwai Smith notes in *Decolonizing Methodologies*.[38] An academic disavowal of the spiritual may be bound up in these colonial knowledge regimes—a rejection and suppression of the epistemologies, cosmologies, and ways of life of many peoples around the globe.

The fallacy of many secularized academic arguments is that they make ethno-nationalist religious regimes a stand-in for all spiritualties, understanding religion as only hegemonic. This stance denies forms of religious oppression, as it also marginalizes the cosmologies of peoples all over the world. It is a false equivalence to read Indigenous cosmologies and spiritualties in the same terms as the faith-based traditions that have been weaponized by the Right. A considered study of cosmo-epistemologies would be inclusive of spiritualties and religions. This has significant import for the environmental humanities which enters questions of futurity and history and humanity's place on the planet and in the cosmos. Indeed, as Anna Gade notes in her study of Muslim environmentalisms:

> both the method and substance of the study of religion (and specifically in regard to the study of Islam) challenge environmental humanities to articulate legibly and systematically its claims about ontology, precarity, nihilism, and so on, as they are increasingly promoted in an existential key.[39]

While the bracketing of religion should be questioned, it also remains true that religious formations have been a mechanism of nation-building and colonial violence. This legacy remains and is carried out in overt and insidious ways: The appropriations of white eco-spiritualism are an example of how spirituality may be a site of epistemic violence. Indeed, white eco-spiritualism has been culpable in what bell hooks labeled "eating the other," and what Wendy Rose (Hopi) later called "white shamanism," referring to the white misuse of culture that Black and Indigenous people create and sustain.[40]

Mindfulness and anti-oppression

Despite the risks and limitations described earlier, a large number of teachers and writers engage in the relationship between anti-oppressive practice and contemplative practice.[41] These theorists have demonstrated how mindfulness practices can be supportive of anti-oppression and anti-racist work, both in and out of the classroom.

As Rhonda Magee writes of the role of mindfulness,

> personal awareness practices are essential for racial justice work. In order for real change to occur, we must be able to examine our own experiences, discover the "situated" nature of our perspectives, and understand the ways in which race and racism are mere cultural constructions

that have a dire impact.[42] Putting her message in the means, Magee's book *The Inner Life of Racial Justice* includes "micro-reflections" that enable readers to practice mindfulness in the moment of reading, calling the reader to become more aware of their experience. "What was it like for you to read those words?," Magee writes, "What thoughts are coming up for you? What do you feel in your body now? What emotions do you feel as you consider what you just read?"[43] These moments are written from the premise that "mindfulness is essential to developing the capacity to *respond*, rather than simply *react* as if on autopilot to what we experience."[44] The distance that mindfulness creates between the fleeting thoughts of the mind and our experience of them is critical to moving from pre-programmed reaction to considered response. The pre-programmed reaction is prone to manifest the dominant or status-quo. Magee's book demonstrates how "racial literacy requires emotional awareness."[45] Magee finds mindfulness to be useful because it can be a way for the white person to move through crises of thinking (as a white racial frame breaks open) in a way that stays with the breath, practicing loving-kindness that does not shore up the ego. This is a radically different approach than behavior modification and rhetorical correction of white ego-defenses (i.e., approaches that seek to curb a list of white behaviors such as credentialing, virtue signaling, ally theater performance, defensiveness, fragility, benevolence, saviorism, and so on). Mindfulness, for Magee, is a resource for trauma-resilience, for healing, for processing racial discomfort, and for coming to consciousness.

The work of dismantling racism is an ongoing, deeply responsive practice, a life's work. "[R]acism is the air we breathe," Jasmine Syedullah reminds in *Radical Dharma*. "There's no getting out of being implicated by it. We're all affected. Acknowledging it is part of what we can do to fight it."[46] Pausing to notice the physical breath may be one way in which we can stop the automatic uptake of beliefs that systemically surround us.

Charles Mills notes how these beliefs orient us to the world and, he writes,

> it is a rare individual who can resist this inherited orientation. Once established in the social mind-set, its influence is difficult to escape, since it is not a matter of seeing the phenomenon with the concept discretely attached but rather of seeing things *through* the concept itself.[47]

How can we interrupt the automaticity of this process, in order to dislodge and throw off what has been imposed?

In a brief passage in his book *Pluriversal Politics*, Escobar looks to the full awareness of mindfulness meditation as a "practice that prepares us to destroy . . . conviction[s] that we have always lived with; it allows us to train our minds to overcome . . . obfuscation."[48] The stillness of meditation allows one to notice long enough to detach from what we've been accustomed to see, potentially snapping open the net of dominant epistemologies that may be experienced as an enclosure.

Speaking of her own perceptual experiences observing the Urubamba River, with rituals offered by Q'ero elders, Macarena Gómez-Barris suggests: "If one can slow down and really listen to the bells, or the sound of the river, we might tread outside of the exigencies of colonial fear and the consumer logic of desire toward a more capacious sensitivity."[49] Gómez-Barris points toward knowledges that can't be codified or captured; instead, we call our attention to them, building awareness.

This practice in part entails being present to the iterative question: What does it mean to speak, act, and be with a persistent awareness of all I do not know and a firm insistence upon the redress of power inequities?[50] This is more than a cognitive practice; mindfulness practice Magee writes, is essential for racial justice work. "Because there are so many rivers of pain joining and forming the ocean of racial suffering in our times," she writes,

> personal awareness practices are essential for racial justice work. In order for real change to occur, we must be able to examine our own experiences, discover the "situated" nature of our perspectives, and understand the ways in which race and racism are mere cultural constructions.[51]

Owen-Smith makes a similar point, describing mindfulness "as a witnessing of our subjectivity that leads to an interruption and transcendence of our habitual perspectives and a freedom from our inflexible thought patterns."[52] She continues, "we cannot consider, interrupt, and challenge oppressive systems until we are fully aware of them, pay attention to them, and reflect on them with a clarity of consciousness."[53]

The self-reflexive work described here is premised on the idea that forms of oppressions are enacted through us. Oppressive ideologies are enacted through everyday perception, decision-making, regard, and action. "Microaggression" is a term that provides one way of naming this reality, but overlapping systems exist in and between us in ways that exceed any list of singular occurrences. As Beth Berila notes,

> Oppressive ideologies insinuate themselves into our very selves, which means that they inevitably inform [our] mental chatter. . . . It takes the capacity to recognize the negative mental chatter and the skills to interrupt it if we are to truly unearth the seeds of oppression.[54]

The "mental chatter" that Berila refers to is akin to what Dianne Prevatt-Hyles and Jana Vinsky call "systemic chatter," a term used to describe how the systemic plays out in internal dialogues and daily enacted patterns.[55] The narratives we receive from outside teach us who we are in relation to one another; interrupting these narratives is necessary in order to find a different configuration and in order to dismantle hierarchical systems that are sustained through us.

To this conversation, Jean Nienkamp offers the concept of "internal rhetoric" to describe what "shapes the rhetorical self."[56] Nienkamp notes, internal rhetoric "is typically a conservative psychic mechanism, adhering most to the deeply ingrained voices." Changing these internal rhetorics takes time and focus: Nienkamp writes that it often takes "years (if ever) to internalize or adhere to new voices that are called forth when people enter new social situations or discourses."[57] Nienkamp here points to the sociality of the mind—the ways in which the mind is given over to other voices and lexicons it encounters in the world. Affective-thought constellations represent a throughline between the individual and the systemic.

One may be predisposed to read the world through scripts and understandings that are culturally provided. In turn, there is the potential to either perpetuate or disrupt (to various extents) these scripts through discourse and action. By mindfully slowing down the "automaticity"[58] of our thoughts and our processing of the world, we come to notice our internal scripts and how they reproduce dominant structures and forms of oppression in our society. Noticing requires breaking from the automaticity of our perceptions and interpretations—or what John Johnson calls "effortful processing"[59] that notices the scripts and schemas that the mind draws upon to make its reading of the world, in order to break from dominant narratives.

This is a starting place: "If the story lines we repeatedly narrate to ourselves are laced with oppressive ideologies," Berila writes, "then the ability to pause those storylines and even rewrite them can go a long way toward creating a more just world."[60] What we said or believed once is not what we cling to for all time; detachment and fluidity are necessary for embracing the possibility of rethinking, reimagining—to prompt the changes that are most needed. This is mindful attention that we are called to ourselves as educators; it is also mindful attention that we invite students to learn so that they can be more present to each other. We can teach our students to pause, to notice the systemic chatter in their minds that might be ignited by forms of encounter in the classroom (through person-to-person exchanges and those that are mediated by a written page etc.). This allows for attention toward narratives and their operations on the ego, attention toward constellations of belief that the self comes to understand itself through. Cultivating a sense of detachment from the ego identity creates space for thwarting these narratives as they are remobilized in our internal rhetorics. Mindful practices can help to draw our awareness to how our presence, words, and action may be a form of resistance or merely part of a matrix that maintains the status quo.

Equity-mindedness, a term popularized by the Center for Urban Education and other DE&I (diversity, equity, and inclusion) scholars, is a paradigm that invites practitioners to "continually reassess their practices to create change."[61] As an intervention, equity-mindedness provides a lens for understanding each individual's practice in an institutional and systemic context. This entails ceaseless, enduring work; as a moment-by-moment practice—changing one's orientation, our way of thinking, acting, and being in the world. Equity-mindedness, which we might think of as equity-*mindful*ness, means developing the attention to notice, for instance, that this comment on a paper, this choice in the classroom, this response to a colleague (and so on) all need an ongoing meditation on the ways that white supremacist ableist heteropatriarchy and colonial logics get reproduced and reinforced through us. The words "consciousness," "awareness," "attention," and "self-reflection" regularly appear in DE&I literature. Such terms perhaps indicate the potential role that mindfulness can play in equity-minded work. Although the focus of "equity-mindedness" is also on institutional accountability, policy, and data-driven response, equity-mindedness also points toward the role of the individual as a potential agent for change, a potential interruption of the system as it stands.

Limitations of mindful approaches to anti-oppression

Yet there are limitations here. Our language of "equity-mindedness" or "consciousness" does not necessarily translate to resistance. Berila notes how the language of mindfulness can appear in "feminist talk of 'consciousness raising' or 'raising awareness,'" yet this phrasing rarely translates to embodied attentiveness, nor does it necessarily translate to action and prompted change.[62] Often these terms are used to reaffirm the saviorist myth that the white feminist "knows best," having done so-called consciousness-raising work. This is a mastery narrative latent in the discourse.

A certain performativity of mindful orientations can fall into another trope: that of white innocence, which Sherene Razack calls a "fantasy of settler civility"[63] or what Melamed, following James Baldwin, calls the "sentimental discourses of liberal reform (past and present)," with "virtuous rage" as a performative "instrument for white salvation."[64] As Melamed explains, "Liberal multiculturalism socialized whites to see themselves as good antiracists by virtue of their antiracist feeling and desire for diversity, even as whites continued to accrue unearned benefits from material and social arrangements that favored them."[65] White feeling and white awareness do not translate to radical change; indeed, centering white feeling and awareness is incompatible with antiracism. The white person will not escape complicity in continued racism through feeling and knowledge. Tropes of white salvation, or what Tuck and Yang call settler "moves to innocence"[66] abound in academic practice: some of the common tropes being settler nativism (i.e., gestures meant to fantasize Indigeneity and belonging to the land which the settler occupies), love of land (i.e., the notion that *if I love well enough and hard enough, I will find reconciliation*), and appropriation of another's pain (i.e., *my empathy will translate into pain being rightly mine and therefore not of my causing*).[67] These tropes are particularly salient in white environmentalism, and white mindfulness practices are also at great risk of these reductive orientations that only reaffirm existing hierarchies. The ubiquity of these moves to innocence draws our attention to the potential for mindfulness to be one more example of white disavowal.

Alcoff clarifies how this repertoire of innocence is structured into social hierarchies:

> If it is true that most people prefer to think of themselves as moral or at least excusable in their actions, then in unjust societies those in dominant and privileged positions must be able to construct representations of themselves and others to support a fantasyland of moral approbation.[68]

White investments in a system that privileges them—and that, in turn, is predicated on a denial of that privilege—reveal themselves through white rhetorical strategies. Catalogues of the "moves that white people make," such as the inventories offered by Eduardo Bonilla-Silva and Robin DiAngelo,[69] reveal the conversational tropes that over-emphasize individual attitudes as a way of obfuscating needed structural changes. White feelings are never the point. The ego that would read movement-building in terms of its own feeling needs tools that are not readily supplied by dominant epistemologies.

Much has been said about the emotional orientation of white fragility. In diversity, equity, and inclusion (DE&I) conversations, it is often described as an unfavorable behavior that is a hindrance to transformative work. Mitigation of white fragility in these spaces might include exercises that unpack white privilege, that teach white people about the limits and potential harms of default repertoires of credentialing, recentering, defensiveness, etc. These interruptions to common white tropes and accompanying behavior modifications may work to prevent certain forms of harm, but they are not addressed to the root: what Walter Mignolo calls the "ego-politics of knowledge."[70] An approach to addressing white fragility, and other "common patterns of well-intentioned white people,"[71] that focuses on warning against discursive moves ultimately misses the point: that is, this approach does little to support the movements that have long been transformative and revitalizing as counters to capitalist-colonialist systems. Antiracist checklists (describing what to do and what not to do), while instructive as heuristics, risk the myth that antiracism can be "mastered" once and for all. A checklist approach risks a righteous fantasy that we can simply read the right books and show up in the right seminars and say the right things to be "cured" of a white racial lens, white supremacy, and racism. This "curative" myth negates the moment-by-moment work of presence in this world. There is no moment of absolution. The work is an ongoing practice. The white ego can only fantasize its innocence or its arrival at "good"; a better intention is for that ego to seek its own undoing—hence the turn to mindfulness and contemplative epistemologies that encourage egolessness and the ability to notice as things shift and change in one's interiority.

I am of course cautious in making this suggestion, keeping in mind how the dominant finds means of reasserting itself. No doubt, if taken up within a dominant knowledge apparatus, mindfulness practices risk becoming part of the mechanism of reproduction and reinstatement. A trenchant mastery orientation and "spiritual supremacism," a term I adopt from Manuel,[72] may be fixed behind a contemplative practice, entrenched there beneath even gentle attention to breath.

Despite the risks noted earlier, topoi of mindful practice that are shared (in not self-same ways) among multiple traditions may allow one to internally witness the depth work a pursuit of epistemic–material justice calls for. Topoi such as nonattachment, attentiveness, nonduality, and slowed-down processing[73] all lend themselves to the work entailed in epistemic revitalization.

The relationship between contemplative practice and egolessness is one reason for turning to this practice: If the white settler selfhood is embedded within systems of oppression, that ego's existence is tied to the perpetuation of these systems. As dominant epistemologies are broken open, the ego they keep intact may be as well.

Reverend angel Kyodo williams writes this of the white ego, addressing white people directly:

> "Whiteness" is a social ego as void of inherent identity as the personal ego, and you have identified with it as much as your very own name, without being willing to name it. . . . But just as the ego-mind cannot be used to work its way out of its own construct, so too can the Mind of Whiteness not be used to see through the veil of its own construct.[74]

Mindfulness may be a way in which the veil comes into visibility, and it is a way in which the ego can find release from its defenses. The white ego comes undone with no new readymade identity to take on. Turning to mindfulness as an exercise in nongrasping is germane to this moment of the ego's dissipation. In mindfulness, one can remain present to the process, not sublimating it or foreclosing it. This crisis of knowing and the crisis of self can be allowed to be, when held in a practice.

In the same text that is co-authored with Reverend angel Kyodo williams, Jasmine Syedullah calls for the abolition of whiteness, at least whiteness as we have known it, because whiteness is an obstruction to abolitionist politics. The habituation to the supremacy of whiteness is a "road to nowhere," "a captivity narrative."[75] White supremacy brings about attachments to normative "perfection," to fear of making mistakes, to exceptionalism, and to status. She also calls out the subtly toxic attributes of whiteness that require "tiptoeing around race and other forms of difference as if in fear of waking a sleeping lion."[76] Waking up to these conditions means, "It is going to get messy."[77] But "the deep connection between liberation and social transformation is increasingly clear. It is embodied," Syedullah writes, with her co-authors williams and Lama Rod Owens.[78]

In an interview in the book, Lama Rod Owens continues this thread regarding the importance of internal work, as not separate from broader transformation. He observes,

> Many of us are really unstudied, unexamined, unquestioned. We're kind of operating on the sense of what we think we're talking about. We're not

> located within our experiences; we're not embodied. Because that really takes opening the closet and owning what's in there and being vulnerable around that. . . . [T]here's still this work of healing that needs to be done for everyone, and we need to bring more attention to that piece. . . . I'm working to end racism and oppression, but at the same time I want to be liberated. I want to thrive. I want to be happy.[79]

Also calling for the attention that Owens speaks of, Zen Buddhist Priest Zenju Earthlyn Manuel instructs, "In order to forget the self, we must study it. We must look at the identities that the self is emotionally attached to."[80] Contemplative practice can enable this kind of study, permitting practitioners to notice mental, emotional, and material habits that support ongoing oppression.

This study includes finding the means of noticing what Claudia Rankine and Beth Loffreda call "the structuring of our very feelings and thoughts by what preceded us." This structuring "is not our own, yet conditions our experience nonetheless," they write.[81] Finding the ability to notice this conditioning in our interiority is an ongoing practice; it cannot be simply "achieved." It shows up anew in each encounter and moment of lived experience.

"Just as anti-oppression pedagogy teaches how to recognize patterns in society," Berila writes of contemplative practice and anti-oppression work, "mindfulness teaches us how to recognize patterns in ourselves," patterns that play out through our daily practice.[82] The act of pausing can help one to notice ideological closures and identify how the mind and body may be conditioned by dominant narratives and assumptions. In the act of pausing and finding stillness, the mind can find even a small amount of distance from the epistemologies that shape one's knowing, coming to recognize the contingency of a thought, separating from it, and creating space for the otherwise. As M. Jacqui Alexander explains in *Pedagogies of Crossing*, privilege means "believing the official story"[83]; a mindful practice can create separation from that belief—just in the act of noticing that a story has won one's assent. This act of noticing is cultivated over time, as a practice.

Contemplative practice is also a resource for what Angel Acosta calls the "prosocial triad": including cultural humility, fierce compassion, and discomfort resilience.[84] The latter is particularly important[85]: the work of destabilizing existing knowledge structures in the pursuit of epistemic justice will cause discomfort (e.g., fear as mental scripts are called into question, anxiety at the encounter of the new, re-lived traumas, and any number of other experiences). Acosta creates healing-centered spaces that develop a diverse group of people's capacity to be vulnerable and present to each other in their varying forms of discomfort, in order to create different circumstances for and with each other. These healing-centered spaces are characterized by mindfulness.

We can, through contemplative practice,[86] cultivate a posture that makes us real to one other—that fosters connection, trust, and what Yuria Celidwen calls "coming close" in respect and without taking. "This means never achieving and always becoming," Celidwen tells us: "This is a committed process." This is listening that undoes the enforcements of personal shame, as it also undoes the systems that allow an epistemology of ignorance to continue. Mindfulness is a continual process, and it is useful for that reason. We don't someday "arrive" at a state of achievement, having "accomplished mindfulness" (the phrase is oxymoronic). Because it resists mastery and because it means dwelling at the roots of our existence, mindfulness may be lent to the radical.[87]

Mindful attention is more than reflection. It breaks out of the discursive-conceptual mind, which, as Wong notes, has limits. Wong elucidates these limits in this way: Explaining of the mind as constructed in dominant systems:

> its binary framing of oppression and anti-oppression implies the moral categories of 'bad' and 'good.' Such erroneous dualistic conceptions of oppression and anti-oppression allow those who identify as anti-oppressive (and morally 'good') or as the oppressed to claim innocence and to avoid examining their own implications in oppressions along the multiple axes of power. It can also result in those who are deemed privileged becoming defensive, frozen, or guilt-stricken. At worst, students may become further separated from each other in their constructed identities instead of coming together to work toward an inclusive and socially just world.[88]

Mindful attention exceeds propositional discourse. Some of the observances of the interior world cannot be codified in language and thus may have the potential for escaping categorical thinking.

Along similar lines, also addressing the limitations of good/bad binaristic thinking, Reverend angel Kyodo williams writes,

> For too long we've been beholden to a set of surface feelings, organizing around ideas and beliefs about what it means to be a good person or create good society. These efforts at good behavior and pursuit of good policies have proven to be no match for the deep embeddedness of what is the foundation of, and has been intricately woven throughout, every face, institution, and relationship of the United States and its inhabitants: the racialization of people and its underlying presupposition—the superiority of white-skinned peoples.[89]

Mindfulness provides a way of breaking out of this limited moralizing thinking. It reaches deeper.

The deep embeddedness of white supremacy is sustained by a web of storylines that need to be unraveled. Reverend angel Kyodo williams notes that white supremacy is a "massive entanglement" that is "in every aspect of how we think, feel, dream, and act toward ourselves and others based on our perception of . . . the social order."[90] She calls for an investigation of the unsatisfactoriness of what has become rooted in our own minds as we confront the ways in which we are individual expressions of a system that has produced great suffering and violence, hindering autonomy, dignity, and thriving.[91] We might say that forms of resistance may be planted in internal investigation, facilitated by stillness. Stillness may activate greater clarity and subsequent courageous action, playing one's part in movement-building while leaving behind the concerns of the ego.

On stillness

Several writers who grapple with the intersections of antiracist practice and mindfulness in the classroom call for the need to make room for silence and stillness in teaching practice. Berila, in her book on contemplative pedagogy, suggests pausing and breathing between each comment made in a class: "Rather than the usual non-stop flow of a class discussion, implement a 'pause and breathe' pattern." The facilitator in the space invites participants "to listen to the speaker, then turn inward, breathe, and reflect on what the person said" and how the expression may have activated certain scripts in the body and mind.[92]

Also theorizing the place of stillness in mindful pedagogy, Owen-Smith importantly identifies how stillness is cultural, and different cultural regard for stillness is too rarely accounted for, especially in educational spaces. In US contexts, Owen-Smith notes, "Silence is often treated as a form of student resistance, oral incompetence, or boredom and, therefore, a response to be remedied."[93] This is one of innumerable examples of how racism, deficit-based thinking, and failure to approach students with cultural humility renders the classroom an exclusionary space. Owen-Smith cites Patrocinio Schweickar who observes that for Filipinos silence is emblematic of cognitive activity and agency; silence "attends wisdom."[94] Referencing Ohiyesa (Charles Alexander Eastman), Owen-Smith also notes how, for many First Nations peoples, silence is "the single most important venue for learning and the sign of perfect equilibrium . . . the absolute balance of body, mind and spirit."[95] Moreover, in his study of silence in Japanese classrooms, Dat Bao found that silence is valued collectively in the Japanese academic system.[96]

In stillness, we have the potential to question the dominance of eurocentric notions regarding silence. To attend to the quiet processes of the mind-in-openness is to "shift the attention from the enunciated to the enunciation," and, in Mignolo's words, by so doing, turn "Descartes' dictum inside out." Being comes before thinking. This mode asks: "what is the quality of being in the mind?" Mignolo sees this moment of inversion of received knowledge as a part of the work of being "epistemically disobedient" to the systems as they stand.[97]

Mindful approaches find the means of renewal in the moment, detaching from and revising our thinking. This happens in the pause, in stillness. Silence, in its physicality, draws us to the body—our body among bodies on land. Finding the multidimensionality of our bodies may enable the process of unlearning the primacy of certain configurations of an intellectual mind and its inheritances.[98]

On affect and emotion

The term "sentipensante" is invoked in relation to a mindful practice, grounded in the inseparability of thinking and sensing. The term "sentipensar" (and its correlate noun, "sentipensamiento") Arturo Escobar notes, are terms "currently used by activists in various parts of Latin America, to suggest a way of knowing that does not separate thinking from feeling, reason from emotion, knowledge from caring."[99] Laura Rendón's sentipensante pedagogy is based on "wholeness and consonance, respecting the harmonious rhythm between the outer experience of intellectualism . . . and the inner dimension of insight, emotion, and awareness."[100] Her work infuses spirit and spirituality into higher education in ways that have been prohibited by the colonial roots of education, which have emphasized a Christian secularism. Rendón seeks a pedagogy that "honors our humanity, instills a sense of wonder, sacredness and humility in our college classrooms, respects and embraces alternate cultural realities, involves social change and healing, and connects faculty and students in meaningful ways."[101] This pedagogy works to counter the privileging of output over self-examination, as it also resists the idea that the same outcomes work for all and that content is interchangeable—that it is appropriate to ask anyone to learn anything.

Tiffany Lethabo King reflects on the emotional toll of this educational model, the ways it is enforced, and how difficult it is to refuse. A sense of "exasperation or justified disgust" that one may feel "because we are not allowed to stop, look at, and more importantly feel the violence of western turns in critical theory" is one of the system's prohibitions. King writes,

> Because of academic respectability politics that impose a kind of bourgeois politesse on all 'communicative acts,' be they in-person or in writing, it is impolite and more importantly irrational to be rendered devastated, enraged, mute, or immobile by the violent terms on which continental theory proceeds. . . . The neoliberal temporality of productivity also requires that scholars keep moving unaffected in the midst of the violence.[102]

King's words expose many of the theories deployed in academic spaces, as they are also an invitation to pay close attention to the affective experience of education. Ensuring our classrooms (whether virtual or physical) do not compress or negate emotion is part of a contemplative orientation. The emphasis on stillness is not meant to translate to a moratorium on anticolonial rage, nor is the emphasis on mindfulness meant to suggest a goal of seeking pleasant sensations and feelings. Mindfulness is not about pleasantness; it is an encounter with what is: in our full embodiment, accompanied by the means to experience otherwise. Mindfulness means staying with experience in its fullness.

The theories of mindfulness presented here, primarily forwarded by BIPOC theorists and practitioners of meditation, mindfulness, and other forms of contemplative practice, honor the mind–body, and counter a legacy in which mind–bodies have been denied in the continued legacy of dehumanization. Affect and embodiment have never been outside of our educational spaces; they are always with us because they *are us*. But a homogenizing rational-emotional habitus is often enforced. Kuokkanen observes a dominant affective orientation of cynicism and fatalism in academic spaces, buttressed by a culture of critique and insistence on certain modes of rationalism. Kuokkanen writes,

> [T]oday's cynicism/ fatalism is very much a product and manifestation of *neoliberal ideology*, which insists that there are no alternatives and that nothing can be done to change the world's present course. . . . Cynicism is a potent mechanism of control that reflects the logic of imperial rationalism. It can be employed as a weapon for dismissing entire worldviews, knowledge systems, and practices as unworthy of serious consideration—as primitive, archaic, simplistic, and gullible.[103]

This fatalist orientation has been noted in environmental humanities scholarship: in the "existential toolkit" project facilitated by Jennifer Atkinson and Sarah Jaquette Ray, in Ray's book *A Field Guide to Climate Anxiety*, in work on emotional resiliency (Leslie Davenport), on sustainable happiness (Greta Gaard), climate fear (Jennifer Ladino), ecological emotions (Panu Pikhala), and many more. Many of these discussions of affect and embodiment in the environmental humanities already draw upon contemplative practice, along with trauma-informed practice. These theorists help break out of the scripts of the classroom and scholarly conventions regarding what can/can't be said, offering the means by which we can draw our attention to the body's experience on the planet. This scholarship pushes back on the institutionally enforced "message that higher education is more interested in . . . intellect than any other aspect of . . . being."[104]

Trauma-informed practice

The theorists named earlier also recognize that trauma is a condition of our climate-changed and violent world—a condition that we are positioned differently in relation to, based on the extent to which we benefit from, or are harmed by, systems of white-supremacist cis-heteropatriarchal capitalist-colonial oppression resulting in a climate-changed world, among many other effects. Given these conditions, antiracist pedagogy necessitates trauma-informed pedagogy.

Antiracist work begins with encountering and being in the presence of longstanding pain, trauma, and suffering—and not denying this pain. "We must never look away," as Ta-Nehisi Coates, in *Between the World and Me*, implores. He writes:

> But all our phrasing—race relations, racial chasm, racial justice, racial profiling, white privilege, even white supremacy—serves to obscure that racism is a visceral experience, that it dislodges brains, blocks airways, rips muscle, extracts organs, cracks bones, breaks teeth. You must never look away from

> this. You must always remember that the sociology, the history, the economics, the graphs, the charts, the regressions all land, with great violence, upon the body.[105]

Privilege is the ability to deny this pain or to merely empathize with it, rather than experience it viscerally day to day. Privilege allows any number of ways of circumventing this rasping reality. Denial happens in the body—as does its converse: attention, not looking away. What one can attend to without falling into voyeurism, performative guilt or innocence, or any other consequence of the distorted ego—this is the form of mindfulness that is needed for the person ready to witness and create change.

A trauma-informed practice, as a necessary part of antiracist pedagogy, recognizes that trauma is a generalized and differentiated condition, not an individualized pathology.[106] Trauma is intergenerational, epigenetic, and pervasive given the conditions of this planet. Trauma is a condition of a racist and oppressive world; it is the inheritance of colonialism, as known from scholarship on Postcolonial Stress Disorder (or Historical Trauma) (Yellow Bird, 2016) and Post Traumatic Slave Syndrome (DeGruy, 2017).[107] These traumas condition the body (Van der Kolk, 2015; Menakem, 2017), and learning is an embodied experience. Thus, any learning environment has trauma in it, even if it goes unacknowledged. The pedagogy that doesn't acknowledge trauma is likely to reproduce it. As Sheila Batacharya and Yuk-Lin Renita Wong write in the introduction to *Sharing Breath*:

> The internal legacies of colonization—the histories of pain and oppression, as well as the guilt-ridden memories of the perpetration of injustice—have been stored up in the body, and the process of recovering and releasing them can be overwhelming and can often take tremendous toll despite the benefits of remedy and repair.[108]

Wong continues in her article in the same book,

> Over the years, it has thus become increasingly crucial for me to support students in learning ways to engage with their emotional reactions to power or holding with gentleness their rage or their wounds from colonization and social injustice, which might be reopened in class discussion.[109]

A trauma-informed contemplative pedagogy is attentive to the ways in which trauma can inhibit one's ability to perceive the body. Fragmentation, hypersensitivity, shame, overwhelming emotion may also be trauma response and facilitators should be conscious of these possible experiences, helping practitioners to know how to prepare for them.[110]

What Wong describes as a trauma-informed pedagogy, Shawn Ginwright usefully suggests as a healing-centered pedagogy. This reframing turns the focus from deficiency and toward the possibility of healing[111]—what is, in his terms, an "explicitly political, rather than clinical" and "culturally grounded," practice of "healing as the restoration of identity" that is guided by participants' own self-knowledge and agency. A healing-centered and healing justice-oriented approach to teaching transforms educational spaces that have historically crushed the soul, as Mary Rose O'Reilly notes; Bettina Love identifies this phenomenon in sharper terms: the "spirit murdering" of BIPOC students.[112]

Mindful practices have a role in healing from the traumas of racism and systemic oppression.[113] Kevin Quashie's book *The Sovereignty of Quiet* celebrates a tradition of Black literature that demonstrates quiet resilience, creativity, healing, and strength. Prioritizing the interiority, Quashie's book counters the racist "determination to see blackness only through a social public lens, as if there were no inner life"[114] through an explication of literature that "asserts the right of a human being to be just that—a human being," an assertion that "privileges the inner life. And yet the interiority of oneness does not correlate with being immune to or isolated from the social world."[115]

As Kirby Moss said in a recent talk on Black joy: "sometimes, instead of moving toward the trauma, I sit out, walk near the ocean, and say 'not today.'"[116] This is a radical presence and not an escape. This embrace of being in the world constitutes an act of remembering who one is, a way of reclaiming one's existence under conditions that relentlessly deny it.

Decomposing the possessive ego: on interrelatedness

In this world, it may be a radical ontological act to reclaim a sense of *being* as a creature interrelated with every other creature on this earth. This is a particular kind of presence, an embrace of the "body that by its very materiality is relational."[117] Pausing to experience that fundamental interconnectedness calls the individualist ego into question, as the ontology invoked in this practice is incompatible with an ontology of separateness. These practices—while commonly undervalued in an education culture that prioritizes disconnected critique and mastery—are essential forms of technê for environmental humanists, which is not to say that they're instrumentalized. The product-driven, categorical, and unresponsive dominant orientation of academe has an imperative of surveillance and the maintenance of the systems by measuring the same inputs and outputs (e.g., learning outcomes) again and again so these become reaffirmed. Transforming these epistemological assumptions requires the cultivation of different ways of being in educational spaces.

AnaLouise Keating, quoting Gloria Anzaldúa, makes an onto-epistemic intervention in this status quo through what she calls the "connectionist" faculty: to show the deep common ground and interwoven kinship among all things and people. This faculty, "one of less-structured thoughts, less-rigid categorizations, and thinner boundaries, allows us to picture—"via reverie, dreaming, and artistic creativity—similarities instead of solid divisions." This is a coalitional act, in Keating's theory.[118] Practicing the connectionist orientation that she describes, Keating draws lines across a wide range of sources and traditions, noting that

> The concept of interrelatedness is a key tenet of many Indigenous worldviews—ranging from the Lakota belief expressed in the phrase "all my relatives" (*mitakuye owasin*), which reminds us that we are related to all existence, to the Buddhist teaching of co-dependent arising, to Thích Nhất Hạhn's theory and practice of inter-being.[119]

Drawing connections between Ubuntu "I am because we are" and Mayan philosophy's "In Lak Ech—yo soy tú otro yo. I am your other I," Keating calls for attention to the radical interdependence of everything in existence.[120] Keating's post-oppositional politics of change emphasizes three main areas: (1) making connections through differences, (2) positing radical interrelatedness, and (3) listening with raw openness.[121]

This theory of interrelatedness counters systems structured to make some beings disposable. Valuing the indispensability of all beings means seeing all more-than-human and human species as interconnected and interdependent. To be interrelated does not mean that we are all the same. We experience the world differently, based on how we are positioned within it and the histories that are particular to our social identities.

Foregrounding interrelatedness counters a capitalist "McMindful" orientation that only shores up possessive individualism. Foregrounding interrelatedness translates mindfulness into an ontological practice that eschews neoliberalism's egoistic, consumerist, and self-enterprising constructions of individualism.[122] This orientation avoids turning mindfulness into a disciplinary technology of the self.[123]

Mindfulness is no panacea, however; indeed, it calls into question the search for a panacea in the first place. None of what is presented here is meant to cover over or absolve conflict. To seek conflict-less spaces is to seek homogeneity. Even within the most seemingly homogeneous communities, there is still difference, and we have conflicts within ourselves. What's more, the suggestion of mindfulness here is meant to disorient an instantiation of the one and known—as part of the work of continually opening space for epistemic justice, which must translate into material change: reparations, the return of lands to Indigenous peoples, respect for the sacred, transformative shifting of power-holding, and continued accountability. I am cognizant that presenting a suggestion such a this risks falling into the trappings of repressive optimism[124] and a failure to grapple with the depth of the harms, the impossibility of solution. In response to the depth effects of violences perpetuated on this globe, heightened awareness alone is no solution. As bell hooks warns in *Communion*, "heightened awareness often gives the illusion that a problem is lessening. This is most often not the case. It may mean simply that a problem has become so widespread it can no longer remain hidden or be ignored."[125]

Classroom practice

What then does a mindful classroom practice look like? It means encountering what occurs in the body in the moment of learning, rather than practicing a sanctioned ignorance of embodiment in a Cartesian space. It means being able to notice and allow sensations, emotions, and thoughts to occur; it means not requiring that these be left at the door. It means inviting ways of naming, but not attaching to, reactions. A thought, a feeling pops up within us; it is impermanent. A reaction does not define who we are. We can choose what we do with the reaction. *What do you want to do with what you just encountered?* We might ask in response to a reading or a conversation in the classroom.

Rhonda Magee recommends naming moments of "racial discomfort." To articulate that "this is a moment of racial discomfort" and that "such moments are common in a world shaped by racism" may be necessary to preventing shame, defensiveness, stonewalling, or other less-than-fruitful reactions. Magee follows the act of naming with a gesture of connection: we can say, "I deserve kindness in this moment. And I offer kindness to others impacted by this moment as well."[126] Such loving-kindness is, for Magee, the ground of accountability. Accountability happens in relationality.

Often moments of racialized (or gendered or xenophobic or otherwise) discomfort are met with avoidance. This avoidance might come in part from denial, shame, or fear: we lack the linguistic and emotional resources for staying with what has occurred. It is beneficial to build a shared lexicon and shared practices for engaging what is always in the room with us: racism, ableism, cisheterosexism, and xenophobia are always there. We are positioned differently from each other within systems that are governed by these axes of oppression. Our identities are therefore always at stake. The collective space of the classroom requires that we find ways of meeting these realities.

We can start with some shared principles, however provisional they might be: I often ask my students on the first day of class: What do you need from me and from each other in order to make this collective space where you want to come to? What can we agree to believe about each other? What can we agree to do for each other? I often ask students to share anonymously, perhaps on a notecard. I collect the responses in one place. From this expression of needs and requests, we come to find a set of shared principles for the community we are forging together.[127] These principles often need to be revisited and revised as we respond to what emerges among us. Among the principles I suggest are that we accept that often our initial reaction to something is a script that is "in the air" of our society, one we've taken up, perhaps unconsciously. To understand an initial thought or feeling as given to us from without, rather than as definitive of what is within, can allow for rethinking and change. This idea also creates space between the thought or feeling and the observation of it. The self that observes doesn't need to hold tight to the initial reaction, doesn't need to defend its hold. This can be a way of precluding the arresting, judgmental postures that are quick to lock into their attachments.

To take seriously this idea of detaching from mental scripts may entail a fundamental rethinking of many conventions and practices. Some pedagogical practices derive from an agonistic tradition that prioritizes stance-taking: It is common to teach thesis statements before we teach students how to stay with the questions, to teach argumentation before we teach how to have an embodied practice of openness. These are examples of how classroom spaces often privilege certain formulations of intellectualism.

What would it mean to cultivate alternative pedagogical spaces? Silence may play a key role in this cultivation.[128] The need for silence is not simply a matter of pausing to "collect one's thoughts." Silence can provide a space for encountering the sublime complexity of the moment in ways that transcend propositional discourse, that surpass what can be identified as a specific thought.

Part of what mindfulness offers in the classroom is a means to encounter the diffractions and entanglements that we are imbricated in. This goes beyond reflection as a cognitive practice. Karen Barad, following Donna Haraway, notes the limitations of reflection and reflexivity, which, in Haraway's words "only displaces the same elsewhere."[129] Reflection is not equipped to meet difference, nor is it generative of difference. "[F]or all of the recent emphasis on reflexivity as a critical method of self-positioning," Barad writes, "it remains caught up in geometries of sameness."[130] Reflection, as a cognitive model, can produce an endless loop of self-examination, with few resources to prompt change.

Haraway and Barad offer diffraction as a "narrative, graphic, psychological, spiritual, and political technology for making consequential meanings." Diffraction means "taking responsibility for the fact that our practices matter" and pausing to notice how "the world is materialized differently through different practices."[131] The ethico-onto-epistem-ology of diffraction, in Barad's analysis, is an orientation toward the "detailed patterns of thinking" of the "fine-grained details" of matter as we exist in entanglements. We are emergent in our interactions and intra-actions; noticing this persistent relationality, and how our practices matter within it, requires slow attention, an escape from the pre-given understandings that are handed down from representational epistemologies and reified discourses.

In the environmental humanities classroom, our "topics" for discussion are often forms of damage, displacement, precarity, and violence that are the lived, intergenerational experiences of many students. There needs to be space in the classroom for feeling this to be true. That doesn't mean that anyone has a right to know another's pain, nor does it mean that pain can be distributed via empathy. What it does mean, however, is that we are embodied beings with histories when we come to the classroom space, and we need to do our best to hold all the complexity of what is there. Pausing and noticing what emerges—and how that emergence may be tied to one's history and the ways in which we are positioned differently in relation to the wake and continuance of oppression—is necessary. *What do your reactions have to tell you about your path?* Feeling daunted or guilty; feeling deep grief or fear; feeling ready—however, we name the feeling can teach us something about who we are and what we are here to do. Allowing *what is* to exist without attaching to it allows for movement, allows space for something else, for what can emerge.

May we not ignore, and may we not leave hidden. We must persist, sourcing the roots of resistance that long precede this moment, roots that are the entitlement of descendents of those who have borne the realities of intersectional oppression. Mindfulness may be one resource for us, in each of our different positionings—as we remain conscious of the fact that mindfulness' roots are long and varied. May we continue to acknowledge the diverse ancestries of mindfulness practices and the diverse worldsenses that sustain a pluriverse, the worldsenses that resist continued colonization.

Notes

1 This statement borrows from Judith Butler who writes of our fundamental dependency in these terms: "I am impressionable, given over to the Other in ways that I cannot fully predict or control." Judith Butler, *Precarious Life: The Powers of Mourning and Violence* (Verso, 2006) 46. Butler calls for a more relational understanding of self-in-world that grapples with precarity as not only an existential condition but also a "social condition from which certain clear political demands and principles emerge." Judith Butler, *Frames of War: When Is Life Grievable?* (Verso, 2009) xxv.

2 Karen Michelle Barad, *Meeting the Universe Halfway: Quantum Physics and the Entanglement of Matter and Meaning* (Duke UP, 2007) ix.

3 For this critique, see Jared Loggins and Wendi Muse, "Read This, Not *White Fragility*," *Dig Podcast*, Jacobin Radio. See also: John McWhorter, "The Dehumanizing Condescension of *White Fragility*," *Atlantic*, 15 July 2020.

4 George Yancy, "Between Pessimism and Optimism: White Crisis," *Humboldt State University*, 10 March 2021, https://president.humboldt.edu/speakerseries/yancy.

5 Resmaa Menakem, *My Grandmother's Hands: Racialized Trauma and the Pathway to Mending Our Hearts and Bodies* (Central Recovery, 2017) ix.

6 Thank you to Evie Ferreira for this theory of the nodal self.

7 The term "technologies of the self" is from Michel Foucault, *The History of Sexuality, Volume 3: The Care of the Self*, translated by Robert Hurley (Vintage, 1986).

8 Jeff Wilson, *Mindful America: The Mutual Transformation of Buddhist Meditation and American Culture* (Oxford UP, 2014) 14.

9 Ronald E. Purser, *McMindfulness: How Mindfulness Became the New Capitalist Spirituality* (Repeater, 2019) 12–13.

10 Purser, *McMindfulness*, 14. These are key points made in critical mindfulness studies (Forbes, 2019; Purser, 2019; Purser, Forbes, Burke, 2016; Wilson, 2014) which, in tandem with other fields and movements such as critical yoga studies (Vats, 2016), interrogates the appropriative and white-washing tendencies of Western mindfulness practices. See David Forbes, *Mindfulness and Its Discontents: Education, Self, and Social Transformation* (Winnipeg: Fernwood, 2019); Ronald E. Purser, *McMindfulness: How Mindfulness Became the New Capitalist Spirituality* (Repeater, 2019); Ronald E. Purser, David Forbes, and Adam Burke, eds., *Handbook of Mindfulness: Culture, Context, and Social Engagement* (Springer, 2016); Jeff Wilson, *Mindful America: The Mutual Transformation of Buddhist Meditation and American Culture* (Oxford UP, 2014); Ajali Vats, "(Dis)owning Bikram: Decolonizing Vernacular and Dewesternizing Restructuring in the Yoga Wars," *Communication and Critical/Cultural Studies* 13.4 (2016): 325–345.

11 Boaventura de Sousa Santos, *Epistemologies of the South: Justice Against Epistemicide*, 2nd edition (Routledge, 2016) 157.

12 Santos, *Epistemologies of the South*, 111. The supple and dynamic act of staying attentive to relationality is what Santos calls an "*ecology of knowledges*," which counter dominant truth.

13 See Daniel P. Barbezat and Mirabai Bush, *Contemplative Practices in Higher Education: Powerful Methods to Transform Teaching and Learning* (Jossey-Bass, 2014); Patricia Owen-Smith, *The Contemplative Mind in the Scholarship of Teaching and Learning* (Indiana UP, 2017); Karen Ragoonaden, ed. *Mindful Teaching and Learning: Developing a Pedagogy of Well-Being* (Lexington Books, 2017).

14 Marie Eaton, Holly J. Hughes, and Jean MacGregor, eds., *Contemplative Approaches to Sustainability in Higher Education* (Routledge, 2017); Greta Gaard, "Mindful New Materialisms: Buddhist Roots for Material Ecocriticism's Flourishing," *Material Ecocriticism*, edited by Serenella Iovino and Serpil Oppermann (Indiana UP, 2014). Jifke Sol and Arjen E. J. Wals, "Strengthening Ecological Mindfulness through Hybrid Learning in Vital Coalitions," *Cultural Studies of Science Education* 10.1 (2015): 203–214; Christine Wamsler, "Mind the Gap: The Role of Mindfulness in Adapting to Increasing Risk and Climate Change," *Sustainability Science* 13 (2018): 1121–1135; Christine Wamsler, Johannes Brossmann, Heidi Hendersson, Rakel Kristjansdottir, Colin McDonald, et al., "Mindfulness in Sustainability Science, Practice, and Teaching," *Sustainability Science* 13.1 (2018): 143–162; Christine Wamsler and Ebba Brink, "Mindsets for Sustainability: Exploring the Link Between Mindfulness and Sustainable Climate Adaptation," *Ecological Economics* 15.1 (2018): 55–61.

15 Laura I. Rendón, *Sentipensante (sensing/thinking) Pedagogy: Educating for Wholeness, Social Justice and Liberation* (Stylus, 2009); bell hooks, *Teaching to Transgress: Education as the Practice of Freedom* (Routledge, 1994); adrienne maree brown, *Emergent Strategy: Shaping Change, Changing Worlds* (AK Press, 2017); Ron Miller, *What Are Schools For: Holistic Education in American Culture* (Holistic Education, 1990). As Rendón notes in her survey of contemplative practice, bell hooks's engaged pedagogy focuses on a union of mind, body, and spirit; the inner life of students and teachers; a connection between classroom and life; and empowerment. hooks's pedagogy is influenced by engaged Buddhism.

16 Michael Yellow Bird, "Neurodecolonization: Applying Mindfulness Research in Decolonizing Social Work," *Decolonizing Social Work*, edited by John Coates and Tiani Hetherington (Routledge, 2016) 298, 305.

17 Owen-Smith, *Contemplative Mind*, 24–25.

18 Yuk-Lin Renita Wong, "'Please Call Me by My True Names': A Decolonizing Pedagogy of Mindfulness and Interbeing in Critical Social Work Education," *Sharing Breath: Embodied Learning and Decolonization*, edited by Sheila Batacharya and Yuk-Lin Renita Wong (Athabasca UP, 2018) 272.

19 Rhonda Magee, *The Inner Work of Racial Justice: Healing Ourselves and Transforming Our Communities Through Mindfulness* (Penguin, 2019) 65.

20 See Daniel P. Barbezat and Mirabai Bush, *Contemplative Practices in Higher Education: Powerful Methods to Transform Teaching and Learning* (Jossey-Bass, 2014) 5–6; Edward W. Sarath, "What Next?: Contemplating the Future of Contemplative Education," *Contemplative Learning and Inquiry across Disciplines*, edited by Olen Gunnlaugson, Edward W. Sarath, Charles Scott, and Heesoon Bai (SUNY P, 2014) 362.

21 Magee, *Inner Work*, 2.

22 See, for example, Arthur A. Costa and Bena Kallick, *Habits of Mind: A Developmental Series* (Association for Supervision and Curriculum Development, 2000).

23 "Habits of mind" approaches have been well-circulated in the university. We should question the ways in which these habits are deployed, how they're conceptualized, and how they're evaluated and by whom. Habits of mind such as "managing impulsivity" and "thinking and communicating with clarity and precision" are racially coded and risk forms of abuse through the policing of behaviors: for instance, assumed definitions of "clarity" and "precision" might fail to account for differences between cultural discourses and the ways in which linguistic discrimination is linked with

racial oppression. Moreover, ideas of persistence and "grit" serve to cover over racist systems, neutralizing them as privatizing the overcoming of them as an individual trait (tenacity) that a person does or doesn't come to practice or represent. See Class Trouble, "A Guide to Coded Language," https://classtrouble.club/blogs/resonance-archives/a-guide-to-coded-language-in-education-vol-i.

24 Owen-Smith, *Contemplative Mind*, 109.

25 Shawn Wilson, *Research Is Ceremony: Indigenous Research Methods* (Fernwood, 2008) 8.

26 Linda Tuhiwai Smith, *Decolonizing Methodologies*, 2nd edition (Zed Books, 2012) 26.

27 Beth Berila, *Integrating Mindfulness into Anti-Oppression Pedagogy: Social Justice in Higher Education* (Routledge, 2015) 149.

28 Wilson, *Mindful America*, 14.

29 Wong, "Please Call Me by My True Names," 272.

30 Zenju Earthlyn Manuel, *The Way of Tenderness: Awakening through Race, Sexuality, and Gender* (Wisdom Publications, 2015) 8.

31 Eaton, Hughes, and MacGregor, in their book on *Contemplative Approaches to Sustainability in Higher Education*, note being drawn to practices that arise from Buddhist, Quaker, Sufi, or other religious traditions because they did not find them in their secularized education.

32 Owen-Smith, *Contemplative Mind*, 2. I am cognizant of the use of religious traditions by ethno-nationalist regimes. There is no innocent space out of reach from dominant systems. That said, ontological practices that have potential for destabilizing these regimes may perhaps still be found and perhaps even in the same religious and spiritual practices that have been mobilized for fascist and authoritarian orientations.

33 Kevin Everod Quashie, *The Sovereignty of Quiet: Beyond Resistance in Black Culture* (Rutgers UP, 2012) 24.

34 Helen Verran, "The Politics of Working Cosmologies Together While Keeping Them Separate," *A World of Many Worlds*, edited by Marisol de la Cadena and Mario Blaser (Duke UP, 2018) 113.

35 Quoted in Santos, *Epistemologies of the South*, 204. See also: Sanya Osha, " Kwasi Wiredu and the Problems of Conceptual Decolonization," *Quest* 13.1,2 (1999): 157–164.

36 M. Jacqui Alexander, *Pedagogies of Crossing: Meditations on Feminism, Sexual Politics, Memory, and the Sacred* (Duke UP, 2006) 15. Quoted in Sheila Batacharya and Yuk-Lin Renita Wong, eds., *Sharing Breath: Embodied Learning and Decolonization* (Athabasca UP, 2018) 12.

37 See Sandy Grande, *Red Pedagogy: Native American Social and Political Thought*, 10th Anniversary edition (Rowman & Littlefield, 2015).

38 Smith, *Decolonizing Methodologies*, 1.

39 Anna Gade, *Muslim Environmentalisms* (Columbia UP, 2019) 249.

40 Wendy Rose, "For the White poets who would be Indian," *Lost Copper* (1980). Quoted in Becky Thompson, *Teaching with Tenderness: Toward an Embodied Practice* (U of Illinois P, 2017) 4.

41 See Sheila Batacharya, and Yuk-Lin Renita Wong, eds., *Sharing Breath: Embodied Learning and Decolonization* (Athabasca UP, 2018); Beth Berila, *Integrating Mindfulness into Anti-Oppression Pedagogy: Social Justice in Higher Education* (Routledge, 2015); Ruth King, *Mindful of Race: Transforming Racism from the Inside Out* (Sounds True, 2018); Rhonda Magee, *The Inner Work of Racial Justice: Healing Ourselves and Transforming Our Communities Through Mindfulness* (Penguin, 2019); Zenju Earthlyn Manuel, *The Way of Tenderness: Awakening through Race, Sexuality, and Gender* (Wisdom Publications, 2015); Resmaa Menakem, *My Grandmother's Hands: Racialized Trauma and the Pathway to Mending Our Hearts and Bodies* (Central Recovery P, 2017); Deborah Orr, "The Uses of Mindfulness in Anti-Oppressive Pedagogies: Philosophy and Praxis," *Canadian Journal of Education* 27.4 (2002): 477–490; angel Kyodo williams, Lama Rod Owens, and Jasmine Syedullah. *Radical Dharma: Talking Race, Love, and Liberation* (North Atlantic, 2016).

42 Magee, *Inner Life*, 7.
43 Magee, *Inner Life*, 16.
44 Magee, *Inner Life*, 17.
45 Magee, *Inner Life*, 27.
46 angel Kyodo williams, Lama Rod Owens, and Jasmine Syedullah, *Radical Dharma: Talking Race, Love, and Liberation* (North Atlantic, 2016) 124.
47 Charles Mills, "White Ignorance," *Race and Epistemologies of Ignorance*, edited by Shannon Sullivan and Nancy Tuana (SUNY P, 2007) 27.
48 Arturo Escobar, *Pluriversal Politics: The Real and the Possible* (Duke UP, 2020) 19.
49 Macarena Gómez-Barris, *The Extractive Zone: Social Ecologies and Decolonial Perspectives* (Duke UP, 2017) 65.
50 This means recognizing that I, as a white person, will need to know that I am, and am understood to be, a beneficiary of systems of oppression. My words are heard in these terms. My positionality shapes what I say; my ethos in the communicative moment is informed by my subject-position. I need internal resources in order to understand myself and be embodied with this awareness.
51 Magee, *Inner Life*, 7.
52 Owen-Smith, *Contemplative Mind*, 33.
53 Owen-Smith, *Contemplative Mind*, 37.
54 Berila, *Integrating Mindfulness*, 19–20.
55 Dianne Prevatt-Hyles and Jana Vinsky, "The LPI Reflective Practice Approach: Anti-Oppression (AOP) In Action," *Liberation Practice International (LPI)*, July 2009, www.liberationeducation.com.
56 Thanks to Alexandria Peary for pointing me to this reference at the Rhetoric Society of America conference and in her book *Prolific Moment*.
57 Jean Nienkamp, *Internal Rhetorics: Toward a History and Theory of Self-Persuasion* (Southern Illinois UP, 2001) 129. Nienkamp delineates *primary* internal rhetorics, the internalization of social language practices that shape identities or selves often at a subconscious level, and *cultivated* internal rhetorics, which are more conscious forms of self-talk. Nienkamp clarifies, "Although it is useful to separate primary from cultivated internal rhetoric for analytic purposes, in practice the two are not so easily distinguished" (126).
58 Adam Lueke and Bryan Gibson, "Mindfulness Meditation Reduces Implicit Age and Race Bias: The Role of Reduced Automaticity of Responding," *Social Psychological and Personality Science* 6.3 (2014): 284–291; John Johnson, "Automaticity to Authenticity: Efforts to Prevent Student Misgendering," Presentation at Sacramento State University, Center for Teaching and Learning, 4 May 2018.
59 Johnson, "Automaticity to Authenticity."
60 Berila, *Integrating Mindfulness*, 20.
61 Center for Urban Education, "Equity-Mindedness," https://cue.usc.edu/about/equity/equity-mindedness/
62 Berila, *Integrating Mindfulness*, 44.
63 Sherene Razack, *Dying from Improvement: Inquests and Inquiries into Indigenous Deaths in Custody* (U of Toronto P, 2015) 210.
64 Jodi Melamed, *Represent and Destroy: Rationalizing Violence in the New Racial Capitalism* (U of Minnesota P, 2011), xi, xii.
65 Melamed, *Represent and Destroy*, 37.
66 Or what Boggs et al. call "the will to epistemic exculpation" or the "constant and obliviating self-absolution" experienced by beneficiaries of predominantly white institutions. See Abigail Boggs, Eli Meyerhoff, Nick Mitchell, and Zach Schwartz-Weinstein, *Abolitionist University Studies: An Invitation* (2019), https://abolition.university/invitation/
67 Eve Tuck and K. Wayne Yang. "Decolonization Is Not a Metaphor," *Decolonization: Indigeneity, Education, & Society* 1.1 (2012): 14 and 15.
68 Linda Martín Alcoff, "Epistemologies of Ignorance: Three Types," *Race and Epistemologies of Ignorance*, edited by Shannon Sullivan and Nancy Tuana (SUNY P, 2007) 49. As

Alcoff notes, this fantasyland includes notions of meritocracy, the idea that modernity began in Europe and then spread outward, and that global poverty is disconnected from Western wealth (49). Alcoff's adoption of the term "fantasyland" comes from Charles Mills discussion of white racial ignorance, a "delusional world, a racial fantasyland, a 'consensual hallucination'" (Mills 18). White practices and beliefs shore up this delusion.

69 Eduardo Bonilla-Silva, *Racism without Racists: Color-Blind Racism and the Persistence of Racial Inequality in America*, 5th edition (Rowman & Littlefield, 2017); Robin DiAngelo, *What Does It Mean to Be White?: Developing White Racial Literacy* (Counterpoints, 2016).

70 Walter D. Mignolo, *The Darker Side of Western Modernity: Global Futures, Decolonial Options* (Duke UP, 2011) 9.

71 DiAngelo, *What Does It Mean*, 199.

72 Manuel, *Way of Tenderness*, 113.

73 Notably, these understandings of contemplative practice are generalized, somewhat dislocated from tradition. This list of topoi does not account for the fact, for instance, that "over the sweep of Asian Buddhist history, regardless of lineage or location," Wilson notes, mindfulness was "presented as a strenuous, lifelong task" that "was decidedly not a process of inhabiting the present moment . . ." nor a practice of "enjoy[ing] the activities of daily life via mindful attitudes." The Buddhist "framework of renunciation and detachment" that gave rise to the Pali term *sati* (the word for what is now prevalently translated into English as "mindfulness") is rarely acknowledged by contemplative pedagogy practitioners of the Anglophone world. Of course, the pursuit of "originary" meanings is fraught. Moreover, the shifts that come with cultural hybridization were, as Wilson notes, Buddhism's history from the start: Buddhism "moves into new cultures and becomes domesticated: in each case, members of the new culture take from Buddhism what they believe will relieve their culture-specific distresses and concerns, in the process spawning new Buddhisms (sometimes, crypto-Buddhisms) that better fit their needs." Wilson, *Mindful America*, 21–22, 3.

74 Williams in Williams, Owen, Syedullah, *Radical Dharma*, xxvii.

75 Syedullah in Williams, Owens, Syedullah, *Radical Dharma*, 23.

76 Syedullah in Williams, Owens, Syedullah, *Radical Dharma*, 12.

77 Syedullah in Williams, Owens, Syedullah, *Radical Dharma*, 21.

78 Syedullah in Williams, Owens, Syedullah, *Radical Dharma*, 39.

79 Owens in Williams, Owens, Syedullah, *Radical Dharma*, 52.

80 Manuel, *Way of Tenderness*, 18.

81 Claudia Rankine and Beth Loffreda, *The Racial Imaginary: Writers on Race in the Life of the Mind* (Fence, 2015) 21.

82 Berila, *Integrating Mindfulness*, 21.

83 Alexander, *Pedagogies of Crossing*, 2.

84 Angel Acosta, "Community Grounding: Contemplating 400 Years of Inequality," Mind & Life Institute, online presentation, 25 June 2020.

85 Lama Rod Owens expounds on this consideration of discomfort in the book *Radical Dharma:* "When I am in tune with my discomfort, I am less likely to avoid your discomfort. Or to put it another way: when I am able to show up to my suffering, I can also show up to yours." Williams, Owens, Syedullah, *Radical Dharma*, 71.

86 Humility is a core to many contemplative practices. Contemplative practice and mindfulness may be a way of developing humility, since humility requires self-understanding and reflexivity. See Arthur Zajonc, *Meditation as Contemplative Inquiry: When Knowing Becomes Love* (Lindisfarne, 2008) 24.

87 I am here alluding to Angela Davis's oft-quoted statement: "Radical simply means 'grasping things at the root.'" Quoted in Walidah Imarisha, Robin D.G. Kelley, and Jonathan Horstmann, "Black Art Matters: A Roundtable on the Black Radical Imagination," *Red Wedge*, 26 July 2016, www.redwedgemagazine.com/online-issue/black-art-matters-roundtable-black-radical-imaginatio.

88 Wong, "Please Call Me by My True Names," 259.

89 Williams in Williams, Owens, Syedullah, *Radical Dharma*, xii.
90 Williams in Williams, Owens, Syedullah, *Radical Dharma*, xiii.
91 Williams in Williams, Owens, Syedullah, *Radical Dharma*, xxiii. On the question of detachment from culturally pre-given ways of thinking and knowing, Alexandria Peary writes in her book *Prolific Moment* of the value of mindfulness practice in enabling "a more detached view of phenomena, including the self. In this detachment, "we are less likely to be tricked into elaborate storylines" (30). She continues, regarding the storylines that are quieted in the moment of meditation: "Sitting on a cushion, a meditation practitioner might be aware of that inner talk for the first time in hours. Previously, self-talk had manipulated the individual, steering their hours, formulating their outlook, interactions, and movements, all without the purview of the person. The person functioned as a sort of sleepwalker, obeying the commands of self-talk, copying its moods, parroting its opinions, almost mechanically" (31). If this "self-talk" is understood as the uptake of dominant ideology, the moment of pausing (on the cushion or in any other moment of daily life) can create the space for detachment.
92 Berila, *Integrating Mindfulness*, 29–30.
93 Owen-Smith, *Contemplative Mind*, 23.
94 Patrocinio Schweickar, "Speech Is Silver, Silence Is Gold: The Asymmetrical Intersubjectivity of Communicative Action," *Knowledge, Difference, and Power: Essays Inspired by Women's Ways of Knowing*, edited by Nancy Rule Goldberger, Jill Mattuck Tarule, Blythe Mcvicker Clinchy, and Mary Field Belenky (Basic Books, 1998). Referenced in Owen-Smith, *Contemplative Mind*, 306.
95 Ohiyesa (Charles Alexander Eastman), "The Power of Silence," *The Soul of an Indian and Other Writings from Ohiyesa*, edited by Kent Nerburn (New World Library, 2001). Referenced in Owen-Smith, *Contemplative Mind*, 7.
96 Dat Bao, *Understanding Silence and Reticence: Ways of Participating in Second Language Acquisition* (Bloomsbury, 2014). Referenced in Owen-Smith, *Contemplative Mind*, 29.
97 Mignolo, *Darker Side*, 160.
98 Here we find resonances with—and space for epistemic and ontological acknowledgment of—the theories that have come with the so-called "ontological turn" in scholarship: "language has been granted too much power," Karen Barad (2003) writes in her work on agential realism, even materiality itself has been transformed into "a matter of language or some other form of cultural representation" (801). Stacy Alaimo, from her chapter in the edited collection *Material Feminism,* writes: "One of the most unfortunate legacies of poststructuralist and postmodern feminism has been the accelerated 'flight from nature' fueled by rigid commitments to social constructionism and the determination to rout out all vestiges of essentialism. Nature, charged as accessory to essentialism, has served as feminism's abject" (237).
99 Escobar, *Pluriversal Politics*, xxxv.
100 Rendón, *Sentipensante*, 2.
101 Rendón, *Sentipensante*, 26.
102 Tiffany Lethabo King, "Humans Involved: Lurking in the Lines of Posthumanist Flight," *Critical Ethnic Studies* 3.1 (Spring 2017): 173–174.
103 Rauna Kuokkanen, *Reshaping the University: Responsibility, Indigenous Epistemes, and the Logic of the Gift* (U of British Columbia P, 2007) 4.
104 Berila, *Integrating Mindfulness*, 38.
105 Ta-Nehisi Coates, *Between the World and Me* (Spiegel & Grau, 2015).
106 Having a trauma-informed lens for pedagogy requires guarding against tendencies to misread overwhelm (especially overwhelm born of trauma) as an unwillingness to learn—readings that easily fall into stereotypes and deficit-based thinking. With a trauma-informed lens, we can stay persistently open to all we don't know (and have no right to know) of an individual's trauma-responses. In general, our orientation should be one that does not label students' behaviours in ways that judge and categorize what we see, turning instead toward an orientation of radical responsiveness to the dynamic human beings we gather with. Trust in the humanity and complexity of what explains

each behaviour, as we also teach students to find this lens in their regard for those who share their educational space and those with whom they coexist on this globe.

107 Michael Yellow Bird writes of neurodecolonization which "makes it possible to overcome oppression, develop the courage to confront it, and cultivate the creativity to strategically challenge it. However, neurodecolonization does not mean that we should suppress or deny negative thoughts and emotions related to colonialism. Yellow Bird, "Neurodecolonization," 299.

108 Batacharya and Wong, *Sharing Breath*, 9.

109 Wong, "Please Call Me by My True Names," 259.

110 See Bessel van der Kolk, *The Body Keeps the Score: Brain, Mind, and Body in the Healing of Trauma* (Penguin, 2015).

111 See also: *Angel Acosta, Healing-Centered Education*, www.drangelacosta.com/healing-centered-education.

112 Mary Rose O'Reilley, *Radical Presence: Teaching as Contemplative Practice* (Heinemann, 1998) 3; Bettina Love, *We Want to Do More Than Survive: Abolitionist Teaching and the Pursuit of Educational Freedom* (Beacon, 2020).

113 See Magee, *The Inner Work* and Menakem, *My Grandmother's Hands.* As Magee asserts, "Mindfulness can help you become more trauma-sensitive and more trauma-resilient" (45).

114 Kevin Quashie, *The Sovereignty of Quiet: Beyond Resistance in Black Culture* (Rutgers UP, 2012) 4.

115 Quashie, *The Sovereignty of Quiet*, 120.

116 Kirby Moss, "On Black Joy," Humboldt State University, 20 February 2020.

117 Christy I. Wenger, "Creating Contemplative Spaces," *Rethinking Ethos: A Feminist Ecological Approach to Rhetoric,* eds. Kathleen J. Ryan, Nancy Myers, and Rebecca Jones (Southern Illinois UP, 2016) 243.

118 AnaLouise Keating, *Transformation Now! Toward a Post-Oppositional Politics of Change* (U of Illinois P, 2013) 567–568. Quoted from Anzaldúa's essay in *This Bridge We Call Home: Radical Visions for Transformation*, edited by Gloria Anzaldúa and AnaLouise Keating (Aunt Lute, 2002).

119 Keating, *Transformation Now*, 46–47.

120 Keating, *Transformation Now*, 71.

121 Keating, *Transformation Now*, 38.

122 As Melamed notes, following Aihwa Ong, self-care discourses can readily translate into technologies of the self meant to render subjects "fit to bear the agency of neoliberal capitalism: egoistic individualism, self-enterprise, and certain calculative practices." Melamed, *Represent and Destroy*, 45.

123 Foucault, *History of Sexuality*.

124 Lauren Berlant, *Cruel Optimism* (Duke UP, 2011).

125 bell hooks, *Communion: The Female Search for Love* (William Morrow, 2002) 116.

126 Magee, *Inner Life*, 29.

127 As an example, one group of students in my environmental writing class came up with the following list in response to this exercise: letting go of perfectionism, self-acceptance, self-forgiveness, knowing that we are understood for our "blindspots" (awareness gaps) and still valued/forgiven by the community, mindfulness about "what is showing up for me," humility and knowing I don't have all the answers, eagerness to learn from each other without demand, knowing that empathetic feedback is a gift shared, a learner's orientation / growth mindset, believing that it's valuable and necessary to be uncomfortable, understanding that cultural ideas are just around us—we receive problematic ideas from our world, but we can change how we think. . . .

128 In her pedagogical practice, Owen-Smith makes silence something that is available for any student to ask for: "The request for a moment of silence at any point during the hour is any class member's call. Often, I will ask for a minute of silence if circumstances warrant, but as the class progresses through the semester, students find their own comfort in calling for silence when they feel the need. . . . Following the silences in the classroom,

we might talk about why silence was needed, what we found from it, and what we might do because of it." Owen-Smith notes the importance of not only auditory silence, but also visual and spatial silence. Owen-Smith, *Contemplative Mind*, 32, 33.

129 Donna J. Haraway, *Modest_Witness@Second_Millennium.FemaleMan_Meets_OncoMouse* (Routledge, 1997).

130 Barad, *Meeting the Universe Halfway*, 72.

131 Barad, *Meeting the Universe Halfway*, 89.

Works cited

Akomolafe, Bayo. "The Times Are Urgent, So Let Us Slow Down." Schumacher College. www.youtube.com/watch?v=bBVAYzBteIo&ab_channel=THISISNOTTHETRUTH

Alaimo, Stacy. *Bodily Natures: Science, Environment, and the Material Self.* Indiana UP, 2010.

Alcoff, Linda Martín. "Epistemologies of Ignorance: Three Types." *Race and Epistemologies of Ignorance*, edited by Shannon Sullivan and Nancy Tuana. SUNY P, 2007.

Alexander, M. Jacqui. *Pedagogies of Crossing: Meditations on Feminism, Sexual Politics, Memory, and the Sacred*. Duke UP, 2006.

Angel Acosta. "Community Grounding: Contemplating 400 Years of Inequality." *Mind & Life Institute*, online presentation. 25 June 2020.

Anzaldúa, Gloria, and AnaLouise Keating, eds. *This Bridge We Call Home: Radical Visions for Transformation*. Aunt Lute, 2002.

Bao, Dat. *Understanding Silence and Reticence: Ways of Participating in Second Language Acquisition*. Bloomsbury, 2014.

Barad, Karen Michelle. *Meeting the Universe Halfway: Quantum Physics and the Entanglement of Matter and Meaning*. Duke UP, 2007.

Barbezat, Daniel P., and Mirabai Bush. *Contemplative Practices in Higher Education: Powerful Methods to Transform Teaching and Learning*. Jossey-Bass, 2014.

Batacharya, Sheila, and Yuk-Lin Renita Wong, eds. *Sharing Breath: Embodied Learning and Decolonization*. Athabasca UP, 2018.

Berila, Beth. *Integrating Mindfulness into Anti-Oppression Pedagogy: Social Justice in Higher Education*. Routledge, 2015.

Berlant, Lauren. *Cruel Optimism*. Duke UP, 2011.

Boggs, Abigail, Eli Meyerhoff, Nick Mitchell, and Zach Schwartz-Weinstein. *Abolitionist University Studies: An Invitation*, 2019. https://abolition.university/invitation/

Bonilla-Silva, Eduardo. *Racism Without Racists: Color-Blind Racism and the Persistence of Racial Inequality in America*. 5th ed. Rowman & Littlefield, 2017.

Butler, Judith. *Precarious Life: The Powers of Mourning and Violence*. Verso, 2006.

Butler, Judith. *Frames of War: When Is Life Grievable?* Verso, 2009.

Celidwen, Yuria. *Indigenous Contemplative Science: An Ethics of Belonging and Reconnection*. Originally broadcast 29 May 2020. The Center for Contemplative Mind and Society. www.youtube.com/watch?v=nJNNKLeB57g&feature=youtu.be

Coates, Ta-Nehisi. *Between the World and Me*. Spiegel & Grau, 2015.

Costa, Arthur A., and Bena Kallick. *Habits of Mind: A Developmental Series*. Association for Supervision and Curriculum Development, 2000.

DeGruy, Joy. *Post Traumatic Slave Syndrome: America's Legacy of Enduring Injury and Healing*. Revised and updated ed. DeGruy Publications, 2017.

DiAngelo, Robin. *What Does It Mean to Be White?: Developing White Racial Literacy*. Counterpoints, 2016.

Eaton, Marie, Holly J. Hughes, and Jean MacGregor, eds. *Contemplative Approaches to Sustainability in Higher Education*. Routledge, 2017.

Escobar, Arturo. *Pluriversal Politics: The Real and the Possible*. Duke UP, 2020.

Forbes, David. *Mindfulness and Its Discontents: Education, Self, and Social Transformation*. Winnipeg: Fernwood, 2019.

Foucault, Michel. *The History of Sexuality, Volume 3: The Care of the Self*, translated by Robert Hurley. Vintage, 1986.

Gaard, Greta. "Mindful New Materialisms: Buddhist Roots for Material Ecocriticism's Flourishing." *Material Ecocriticism*, edited by Serenella Iovino and Serpil Oppermann. Indiana UP, 2014.

Gade, Anna. *Muslim Environmentalisms*. Columbia UP, 2019.

Ginwright, Shawn. "The Future of Healing: Shifting from Trauma Informed Care to Healing Centered Engagement." *Occasional Papers* 25 (2018). Newsletter, accessed on 27 September 2020. http://kinshipcarersvictoria.org/wp-content/uploads/2018/08/OP-Ginwright-S-2018-Future-of-healing-care.pdf

Gómez-Barris, Macarena. *The Extractive Zone: Social Ecologies and Decolonial Perspectives*. Duke UP, 2017.

Grande, Sandy. *Red Pedagogy: Native American Social and Political Thought*. 10th anniversary ed. Rowman and Littlefield, 2015.

Haraway, Donna J. *Modest_Witness@Second_Millennium.FemaleMan_Meets_OncoMouse*. Routledge, 1997.

hooks, bell. *Communion: The Female Search for Love*. 1st ed. W. Morrow, 2002.

Imarisha, Walidah, Robin D.G. Kelley, and Jonathan Horstmann. "Black Art Matters: A Roundtable on the Black Radical Imagination." *Red Wedge*. 26 July 2016.

Johnson, John. "Automaticity to Authenticity: Efforts to Prevent Student Misgendering." Presentation at Sacramento State University, Center for Teaching and Learning. 4 May 2018.

Keating, AnaLouise. *Transformation Now! Toward a Post-Oppositional Politics of Change*. U of Illinois P, 2013.

King, Ruth. *Mindful of Race: Transforming Racism from the Inside Out*. Sounds True, 2018.

King, Tiffany Lethabo. "Humans Involved: Lurking in the Lines of Posthumanist Flight." *Critical Ethnic Studies* 3.1 (Spring 2017): 162–185.

Kuokkanen, Rauna. *Reshaping the University: Responsibility, Indigenous Epistemes, and the Logic of the Gift*. U of British Columbia P, 2007.

Loggins, Jared, and Wendi Muse. "Read This, Not *White Fragility*." *Dig Podcast*, Jacobin Radio.

Love, Bettina. *We Want to Do More Than Survive: Abolitionist Teaching and the Pursuit of Educational Freedom*. Beacon, 2020.

Lueke, Adam, and Bryan Gibson. "Mindfulness Meditation Reduces Implicit Age and Race Bias: The Role of Reduced Automaticity of Responding." *Social Psychological and Personality Science* 6.3 (2014): 284–291.

Magee, Rhonda. *The Inner Work of Racial Justice: Healing Ourselves and Transforming Our Communities Through Mindfulness*. Penguin, 2019.

Manuel, Zenju Earthlyn. *The Way of Tenderness: Awakening through Race, Sexuality, and Gender*. Wisdom Publications, 2015.

McMahan, David L., and Erik Braun, eds. *Meditation, Buddhism, and Science*. Oxford UP, 2017.

McWhorter, John. "The Dehumanizing Condescension of *White Fragility*." *Atlantic*. 15 July 2020. www.theatlantic.com/ideas/archive/2020/07/dehumanizing-condescension-white-fragility/614146/

Melamed, Jodi. *Represent and Destroy: Rationalizing Violence in the New Racial Capitalism*. U of Minnesota P, 2011.

Menakem, Resmaa. *My Grandmother's Hands: Racialized Trauma and the Pathway to Mending Our Hearts and Bodies*. Central Recovery P, 2017.

Mignolo, Walter D. *The Darker Side of Western Modernity: Global Futures, Decolonial Options*. Duke UP, 2011.

Miller, Ron. *What Are Schools For: Holistic Education in American Culture*. Holistic Education, 1990.

Mills, Charles. *The Racial Contract*. Cornell UP, 1997.

Mills, Charles. "White Ignorance." *Race and Epistemologies of Ignorance*, edited by Shannon Sullivan and Nancy Tuana. SUNY P, 2007, pp. 13–38.

Moss, Kirby. "On Black Joy." Humboldt State University. Presentation. 20 Feburary 2020.

Nienkamp, Jean. *Internal Rhetorics: Toward a History and Theory of Self-Persuasion*. Southern Illinois UP, 2001.

Ohiyesa (Charles Alexander Eastman). "The Power of Silence." *The Soul of an Indian and Other Writings from Ohiyesa*, edited by Kent Nerburn. New World Library, 2001.

Ong, Aihwa. *Neoliberalism as Exception: Mutations in Citizenship and Sovereignty*. Duke UP, 2006.

O'Reilley, Mary Rose. *Radical Presence: Teaching as Contemplative Practice*. Heinemann, 1998.

Orr, Deborah. "The Uses of Mindfulness in Anti-Oppressive Pedagogies: Philosophy and Praxis." *Canadian Journal of Education* 27.4 (2002): 477–490.

Osha, Sanya. "Kwasi Wiredu and the Problems of Conceptual Decolonization." *Quest* 13.1–2 (1999): 157–164.

Owen-Smith, Patricia. *The Contemplative Mind in the Scholarship of Teaching and Learning*. Indiana UP, 2017.

Peary, Alexandria. *Prolific Moment: Theory and Practice of Mindfulness for Writing*. Routledge, 2018.

Polinska, Wioleta. "Mindfulness Meditation as a Remedy to 'White Ignorance' and Its Consequences." *Buddhist-Christian Studies* 38 (2018): 325–341.

Prevatt-Hyles, Dianne, and Jana Vinsky. "The LPI Reflective Practice Approach: Anti-Oppression (AOP) In Action." *Liberation Practice International (LPI)*. Toronto, On. Canada. July 2009. www.liberationeducation.com

Purser, Ronald E. *McMindfulness: How Mindfulness Became the New Capitalist Spirituality*. Repeater, 2019.

Purser, Ronald E., David Forbes, and Adam Burke, eds. *Handbook of Mindfulness: Culture, Context, and Social Engagement*. Springer, 2016.

Purser, Ronald E., and David Loy. "Beyond McMindfulness." *Huffington Post*. 31 August 2013. www.huffpost.com/entry/beyond-mcmindfulness_b_3519289

Quashie, Kevin. *The Sovereignty of Quiet: Beyond Resistance in Black Culture*. Rutgers UP, 2012.

Ragoonaden, Karen, ed. *Mindful Teaching and Learning: Developing a Pedagogy of Well-Being*. Lexington Books, 2017.

Rankine, Claudia, and Beth Loffreda. *The Racial Imaginary: Writers on Race and the Life of the Mind*. Fence, 2015.

Ray, Sarah Jacquette. "Coming of Age at the End of the World: The Affective Arc of Undergraduate Environmental Studies Curricula." *Affective Ecocriticism: Emotion, Embodiment, Environment*, edited by Kyle Bladow and Jennifer Ladino. U of Nebraska P, 2018.

Razack, Sherene. *Dying from Improvement: Inquests and Inquiries into Indigenous Deaths in Custody*. U of Toronto P, 2015.

Rendón, Laura I. *Sentipensante (Sensing/Thinking) Pedagogy*. Stylus, 2008.

Rosiek, Jerry Lee, Jimmy Snyder, and Scott L. Pratt. "The New Materialisms and Indigenous Theories of Non-Human Agency: Making the Case for Respectful Anti-Colonial Engagement." *Qualitative Inquiry* 26.3–4 (2020): 331–346.

Santos, Boaventura de Sousa. *Epistemologies of the South: Justice Against Epistemicide*. 2nd ed. Routledge, 2016.

Schweickar, Patrocinio. "Speech Is Silver, Silence Is Gold: The Asymmetrical Intersubjectivity of Communicative Action." *Knowledge, Difference, and Power: Essays Inspired by Women's Ways of Knowing*, edited by Nancy Rule Goldberger, Jill Mattuck Tarule, Blythe Mcvicker Clinchy, and Mary Field Belenky. Basic Books, 1998.

Smith, Linda Tuhiwai. *Decolonizing Methodologies*. 2nd ed. Zed Books, 2012.

Sol, Jifke, and Arjen E. J. Wals. "Strengthening Ecological Mindfulness through Hybrid Learning in Vital Coalitions." *Cultural Studies of Science Education* 10.1 (2015): 203–214.

Thompson, Becky. *Teaching with Tenderness: Toward an Embodied Practice*. U of Illinois P, 2017.

Tuck, Eve, and K. Wayne Yang. "Decolonization Is Not a Metaphor." *Decolonization: Indigeneity, Education, and Society* 1.1 (2012): 1–40.

van der Kolk, Bessel. *The Body Keeps the Score: Brain, Mind, and Body in the Healing of Trauma*. Penguin, 2015.

Vats, Ajali. "(Dis)owning Bikram: Decolonizing Vernacular and Dewesternizing Restructuring in the Yoga Wars." *Communication and Critical/Cultural Studies* 13.4 (2016): 325–345.

Wamsler, Christine. "Mind the Gap: The Role of Mindfulness in Adapting to Increasing Risk and Climate Change." *Sustainability Science* 13 (2018): 11121–1135.

Wamsler, Christine, and Ebba Brink. "Mindsets for Sustainability: Exploring the Link Between Mindfulness and Sustainable Climate Adaptation." *Ecological Economics* 15.1 (2018): 55–61.

Wamsler, Christine, Johannes Brossmann, Heidi Hendersson, Rakel Kristjansdottir, Colin McDonald, et al. "Mindfulness in Sustainability Science, Practice, and Teaching." *Sustainability Science* 13.1 (2018): 143–162.

Wapner, Paul. "Contemplative Environmental Studies: Pedagogy for Self and Planet." *The Journal of Contemplative Inquiry* 3.1 (2016).

Wenger, Christy I. "Creating Contemplative Spaces," *Rethinking Ethos: A Feminist Ecological Approach to Rhetoric,* edited by Kathleen J. Ryan, Nancy Myers, and Rebecca Jones. Southern Illinois UP, 2016, pp. 237–256.

williams, angel Kyodo, Lama Rod Owens, and Jasmine Syedullah. *Radical Dharma: Talking Race, Love, and Liberation*. North Atlantic, 2016.

Wilson, Jeff. *Mindful America: The Mutual Transformation of Buddhist Meditation and American Culture*. Oxford UP, 2014.

Wilson, Shawn. *Research Is Ceremony: Indigenous Research Methods*. Fernwood, 2008.

Wiredu, Kwasi. "Are There Cultural Universals?" *Quest* 4.2 (1990): 5–19.

Wiredu, Kwasi. *Cultural Universals and Particulars: An African Perspective*. Indiana UP, 1996.

Wiredu, Kwasi. "African Philosophy and Inter-Cultural Dialogue." *Quest* 11.1,2 (1997): 29–41.

Wong, Yuk-Lin Renita. "'Please Call Me by My True Names': A Decolonizing Pedagogy of Mindfulness and Interbeing in Critical Social Work Education." *Sharing Breath: Embodied Learning and Decolonization*, edited by Sheila Batacharya and Yuk-Lin Renita Wong. Athabasca UP, 2018, pp. 253–279.

Yancy, George. "Between Pessimism and Optimism: White Crisis." *Recorded Talk*. Humboldt State University. 10 March 2021. https://president.humboldt.edu/speakerseries

Yellow Bird, Michael. "Neurodecolonization: Applying Mindfulness Research in Decolonizing Social Work." *Decolonizing Social Work*, edited by John Coates and Tiani Hetherington. Routledge, 2016, pp. 319–336.

Zajonc, Arthur. *Meditation as Contemplative Inquiry: When Knowing Becomes Love*. Lindisfarne, 2008.

4

READING AND WRITING PRACTICES IN THE CLASSROOM

This chapter attends to some of the literacy practices that might accompany the orientations outlined in the previous chapters. Given that our disciplinary practices include the study and production of texts of various kinds, how can we read and write in alignment with pluriversality? How can reading and writing practices work to counter what Boaventura de Sousa Santos calls the "cognitive empire"[1] that denies the radical pluriversality of lived knowledges?

In this chapter, my focus is primarily on literary reading practices, but most of the discussion pertains to reading practices across the humanities, different genres, and conversations. I choose to focus on literature because I am situated in an English department and literary study traverses the humanities. While a full account of ecocriticism and theories of reading is beyond the scope of this chapter, the question of an environmental humanities literacy practice is of interest here. What we read for, how we frame texts, and what texts make possible are all necessary questions in the environmental humanities.

The potential of literature

Literature is potentially collective making, movement making, Jodi Melamed writes, arguing that literary and artistic works germinated the "roots of radical antiracism."[2] Literature's generative potential comes from its modality and form: for Melamed, literature is a space "in which it is theoretically possible to say anything—that is, a discourse with a bracketed truth imperative." Art can break open sanctioned knowledge and "figure (and sometimes prefigure) materialisms relatively unbound from historically dominant expressions of economic and political value."[3] Reading and writing have the potential to make present what Erin James calls "the world-creating power of narratives."[4] James writes of storyworlding as a fundamentally environmental process, "in which readers come to know what it

DOI: 10.4324/9781003037439-5

is like to experience a space and time different from that of their immediate reading environment."[5] This transport does not necessarily entail a "taking on"—an assumption of belonging to or owning. The process of "coming to know" through reading and writing can give only a glimmer of a world sense—and what it means to inhabit it in its particularity—but literature and art can offer this fleeting fragmentary light. We should resist any notion of literature being "representative" or a "window into" another subjectivity, an "informant" for the reader who has the power to demand transparency. Keeping aware of these risks, the experience of art can snap open routinized ways of sensing the world, in order to perceive the radical contingency of one's own understanding and—especially for those who have the privilege of seeing their inherited worldview everywhere reinforced—the possibility of something else.

The humanities might aspire to stay near those texts that make a departure from hegemonic "truth" possible. But there are limits here: We must "never confuse reading and writing literature with political and socially transformative action," Melamed makes clear.[6] This is both because thought-work does not immediately translate into material change, even as the epistemic is in a continuum with materiality, and because reading practices may domesticate a text and suture it to master discourses. Melamed is observant of the ways in which literatures and the study of them "produce and police national culture."[7] Indeed, "discourses of global literature function within a network of globalized information industries (including universities) to circulate information bits that accord with neoliberal policy and agendas by passing them off as literary knowledge."[8] Relays of knowledge/power may be cloaked behind a belief in art's unique capacity for transformation—and stitched into trust in individual expression. Even in the revolutionary potential of art-making, knowledge/power continues with its arrangements of truth.

Melamed explicates knowledge/power's circuitry, focusing on literary studies in the United States particularly, but in ways that pertain to the education of the professional–managerial class more broadly: university education, she writes, works to socialize students into supposedly "progressive" constituencies that perpetuate a grossly unequal system of global capital accumulation—teaching these students to "perceive themselves as antiracist and multicultural . . . in a manner that allowed the material conditions for a new apartheid between haves and have-nots to flourish."[9] Melamed notes how "[r]acial discipline proceeds from liberal affect,"[10] through the integration of "liberal multicultural concepts into [one's] sense of self-actualization"[11] which amounts to complacency with the system that authorizes the self as being absolved and legitimized through its performance of "a properly multicultural orientation." This performance moves through "discourses of mission, benevolence, and service that are otherwise training students with privilege to play their parts in neoliberalism's civilizing and disqualifying regimes."[12]

What Melamed describes are some of the means by which the humanities classroom works in service of existing stratified orders and their corresponding acts of violence. There is no salve for this reality, and my interest in mindfulness is not meant to suggest it as a remedy. A mindful orientation to pedagogy does not necessarily disrupt the pedagogical replications that Melamed describes. As established in the previous chapter, mindfulness does not produce epistemic justice; instead, we need an onto-epistemic justice orientation to both mindfulness practice and mindfulness-based pedagogy.

Katherine McKittrick in *Dear Science and Other Stories* names a listening practice for texts that may bring forward epistemologies that push up against dominant orders. Citing Dina Georgis, McKittrick writes, "The story. . . . has the capacity to affectively move us and, at the same time, incite a listening practice that is 'neither disengaged nor wanting to master what it sees and hears.'"[13] McKittrick stays with this listening capacity, asking: "What if we read outside ourselves not *for* ourselves but to actively unknow ourselves, to unhinge, and thus come to know each other . . . as collaborators of collective and generous and capacious stories?"[14] This act of reading-to-unknow has the potential to destabilize the knowledges that are enacted through us. McKittrick carves out space for the story's potential. Speaking specifically of a Black literary and theoretical tradition, McKittrick calls story

> a way to hold on to the rebellious methodological work of sharing ideas in an unkind world. Sharing can be uneasy and terrifying, but our stories of black worlds and black ways of being can, in part, breach the heavy weight of dispossession and loss.[15]

McKittrick's faith in the value of literature is echoed across scholarship that focuses on racial and epistemic justice. In the introduction to *The Racial Imaginary*, Claudia Rankine and Beth Loffreda state the importance of literature: "we believe in the beauty and importance of an individual speaking in and to her history with as much depth and seriousness as she can muster, especially when that history doesn't present itself as an otherwise-sayable event."[16] For Rankine and Loffreda, literature offers no space of innocence or removal from continued oppressions, but it can be a way of speaking in counterpower to repeating histories. Creative production—the act of writing, making, and reading—can be the refusal of erasure; it can be an antidote to cynicism and fatalism. Indeed, creative production can break out of received knowledge or the diffuse hegemonies of what is thought to be true: Rankine and Loffreda put it this way: "Our imaginings might test our inheritances, to make way for a time when such inheritances no longer ensnare us."[17] At the same time, the potential of the imaginative work cannot exist as a denial of what is: "we are creatures of this moment," Rankine and Loffreda remind and we must account for as much.

The forms that make up humanistic study allow for the reimagination of our world. It has been said many times by thinkers as ranging as Gloria Anzaldúa and Julia Kristeva: as Kuokkanen reaffirms, "a change in the real is impossible without a simultaneous change in the 'imaginary.'"[18] Rankine and Loffreda say this as well: A societal imaginary "changes over time, in part because artists get into tension with it, challenge it, alter its availabilities."[19]

Myth of the free imagination

However, this transformative potential—this ability to challenge and alter—does not mean that art and literature arrive unburdened by the oppressive regimes of knowing that hang heavy around us. Rankine and Loffreda expose the dangerous falsity of the (white, settler) claim that: the imagination is a free space, and I have the right to imagine from the point of view of anyone I want—and it is against the nature of art itself to put limits on who or what I can imagine. As Rankine and Loffreda note, "White writers often begin from a place where transcendence is a given—one already has access to all, one already is permitted to inhabit all, to address all."[20] This denial of positionality and disavowal of one's fundamental conditioning and embeddedness in power relations is not only born of privilege but is also damaging. It forecloses the means by which one might be called to account for how their thought-work is part of knowledge/power.

Like Rankine and Loffreda, Mohanty also describes the common rhetorical move to fantasize a free, transcendent imaginative–intellectual space. She writes,

> The claim to a lack of identity or positionality is itself based on privilege, on a refusal to accept responsibility for one's implication in actual historical or social relations, on a denial that positionalities exist or that they matter, the denial of one's own history and the claim to a total separation from it.[21]

From Rankine and Loffreda and Mohanty we learn to ask: How is my thinking conditioned by my subject-position and affordances in a world of inequities? How is my thinking not my own but instead bound up with, and made possible by, relations of power? What do I feel entitled to speak of, who do I feel entitled to speak for, and what conditions my sense of authority? What does it mean for me to say what I say, given my identities and my context? How is my ethos always interpellated by the axes of identity that correspond to disparate power (race, gender, citizenship status, etc.)?

These are critical questions that we ask ourselves and those we engage as readers and listeners. To embrace such criticality is congruous with artistic practice. Contrary to the common (white-dominated) tropes that would separate art from critical practice[22] and hold out a "free" imagination as protected from the "anxieties" of such questions, we need to embrace a critical imagination.[23] As Rankine and Loffreda note, to put a critical lens on the imagination—to find its enclosures—is not prohibition, but is instead an "invitation to witness our inevitable racial subjectivity."[24] It is an opportunity "[t]o ask what we think we know, and how we might undermine our own sense of authority,"[25] which is the essence of the epistemic demeanor of humility discussed previously. Undermining one's authority is not to fall into passivity, but instead to stay with what can't be fully known and to not presume a stance of "knowing for."

The myth of the free imagination is one common motif in white discourse, manifested in scholarship that seems to call from nowhere. This neutralized voice from "outside" inhabits what we've understood from Mignolo as a "zero-point" epistemology. That voice which speaks in denial of its situatedness adopts a cloak in authorized knowledge sometimes assumed to be "objective." To counter this zero-point orientation and the ways that it reifies stratified knowledges requires attending to positionality. It means, in Alcoff's words, following Charles Mills, identifying specific knowing practices as they are inculcated and played out in a socially dominant group.

The ubiquity of critique

The myth of the free white imagination is not the only pernicious idea lurking in our reading practices in the humanities. Tema Okun's well-circulated list of white supremacist motifs demonstrates how deeply white epistemologies are embedded in institutional practices. Okun's list emerges from group discussions and note-taking across many workshops, so it represents a collectively authored text that names aspects of white supremacy culture in institutionalized spaces. Okun notes that the 15 major themes identified in the list are "powerful precisely because [they are] so present and at the same time so very difficult to name or identify." They "are used as norms and standards without being proactively named or chosen by the group." These motifs of white supremacy culture reference a pervasive culture of critique: "a tendency to identify what's wrong, [with] little ability to identify, name, and appreciate what's right."[26] How much time do we spend in our classrooms fostering an ability to critique, in comparison with the time we spend enacting celebration and care?

In addressing what is so endemic to institutional culture through our literacy practices in the environmental humanities, we do well to pause on our regard for critique. The ubiquity of critique as a reading strategy has been well-described by Rita Felski in her book *The Limits of Critique*. The book offers ways of naming reading practices in English Studies, but the observations may pertain more broadly across the humanities. Running the gamut between forceful condemnation, troubling, or calling into question, critique is everywhere in academic culture, Felski argues. She asks, "Why does critique hold the charisma it does for the field?" and "Why is it so hard to get outside its orbit?"[27] "Why is critique so frequently feted as the most serious and scrupulous form of thought? What intellectual and imaginative alternatives does it overshadow, obscure, or overrule?" And "what are the costs of such ubiquitous criticality?"[28] "Why—even as we extol multiplicity, difference, hybridity—is the affective range of criticism so limited? Why are we so hyperarticulate about our adversaries and so excruciatingly tongue-tied about our loves?"[29]

Felski names critique as a dominant intellectual reliance. She identifies ways in which critique is embedded in knowledge structures that are divested of radicalism. Felski exposes how critique is structured to be objectifying—always "of" something, pointing the finger at its object, while containing and essentializing it as one thing. Critique is fraught in another way as well, in Felski's analysis: In relation to the object of critique, the critic assumes the posture of the "subject who knows," entitled to diagnose and ameliorate the deficit object of study. Critique is correspondingly always at some level an act of dismissal—while critique as a modality also enables the paternalistic critic to find a space of innocence on the "right side" of the object: That the critic can negate the object of critique without having to take a stance "for" something activates a space of what Felski calls "negative freedom." In turn, the critic ignores their own participation in what is critiqued and enforces an idea that critique (and, with it, the critic) is superior: "Critique chafes at the presence of other forms of thought," Felski writes.[30] This is an assertion of supremacy: the critic authorized to dismiss by carrying out the rhetoricity of critique.

The insistence on modes of critique in the humanities is one way that white epistemologies enclose the texts they encounter. Jessica Lopez Lyman, in an essay titled "Imposter Poet: Recovering from Grad School," writes of how this intellectual insistence plays out in the classroom: "The seminar room teaches you to critique. Through a capitalist, colonial logic most graduate programs destroy texts, decapitate theories, and turn concepts and methods into loose limbs clinging to the page for survival."[31] Lyman here is pointing to a parallel between critical methods and capitalist–colonialist orientations toward destruction-as-progress. Moving the text object through a mechanized process of critique can mean taking it apart, taking from it. This orientation is in contrast to reading in humility, in the presence of another.[32]

Kandace Creel Falcón echoes Lyman's point in the same volume, in an essay titled "What Would Edén Say?," by questioning the notion that:

> "Valid" academic writing and scholarship requires distance, a pretend, yet required, scenario in which the observer is supposed to be outside of that which is being observed. This is laughable. . . . White academia tries to insist the better scholarship is by those who can retain a neutral stance.[33]

This supposed "neutral stance" is that of the dominant identity that is centered, and therefore privileged, as the standard. The texts that go "unmarked," as it were, maintain the invisibility of a white heteropatriarchal order that fantasizes itself to be universal.

What critique means to those who are marginalized by academic systems runs counter to such dominant practice. Tiffany Lethabo King describes womanist theorists who modify Paul Ricoeur's "hermeneutics of suspicion"[34] to describe the "reading and interpretive practices of Black women," highlighting the political power and necessity of distrust toward "heteronormativity or conventional ways of thinking about what is natural and normal."[35]

From King's example, we can see that it is not critical thought that is problematic: it's the normative assumptions of the critic who understands themself to be "outside" and in a position to "know best." It is the tendency to dismiss, to hierarchize, to ignore complicity, and to reinforce supremacist hierarchies. As Boggs et al.

write in their abolitionist manifesto for university studies, "we need to be unsettled by critique's privileged place in the institutional epistemology of the university, in which the status it enjoys as a good in itself is enshrined."[36] They continue by explicating critique's "organizational imaginary that is a means of university reproduction." They write, "The expectation that critique is a sufficient vehicle for the enactment of our politics needs to be counterbalanced with a historical understanding of critique as institutional embeddedness."[37] For those embedded in dominant systems of thinking, the intention is that we might see the narratives that shape our lives—to dissociate with the cultural imagination they inhabit and then to imagine otherwise, without appropriating, fantasizing our identities as other than they are, denying one's place as a beneficiary of a system or failing to leverage that place and position.

The relentlessness of academic critique intersects with several of the characteristics of white supremacy culture outlined by Tema Okun, including perfectionism ("pointing out either how the person or work is inadequate"), paternalism (or knowing better than, knowing for), power hoarding ("those with power assume they have the best interests at heart"), objectivity, and so on.[38] Tiffany Lethabo King identifies the technê of critique as consisting of the following steps: "know, internalize, perform, disagree, and then center yourself."[39] This is an emaciated practice. As Chris Finley notes, "enlightenment provides us with the tools to critique and compare, but not to love and understand each other."[40]

Politics of refusal

Identifying these features of critique is not enough to dismantle it: As King notes, we should have a

> deep distrust of the ability of Western theory—specifically its attempt at self-critique and self-correction in the name of justice for humanity—to revise its cognitive orders to work itself out of its current "closed system," which reproduces exclusion and structural oppositions based on the negation of the other.[41]

An examination of modes of white critique provides just one foothold for tearing down what King calls the "very systems, institutions, and order of knowledge that secure the humanities as an exclusive experience and bound identity in violent ways."[42]

King's words are not an abstraction. As she notes, "We need to consider on whose backs or through whose blood a theory developed and then circulated while hiding its own violence."[43] Or, in Tuck and Yang's words, how does the text—or an approach to it—replicate the triad relationship of settler colonialism:

> between the white settler (who is valued for his leadership and innovative mind), the disappeared Indigenous peoples (whose land is valued, so they and

> their claims to it must be extinguished), and the chattel slaves (whose bodies are valuable but ownable, abusable, and murderable)?[44]

We need to persistently ask the question, as King does, in alignment with Tuck and Yang: "If this line of thought requires Indigenous death, why even venture down it? What could one possibly repair or salvage of it?"[45] This pertains to all the texts we introduce into the classroom: Do they forward the myth of the wilderness? Do they erase violence, genocide, and enslavement? Do they heroicize white people as saviors? Elizabeth Cook-Lynn's well-known essay "Why I Can't Read Wallace Stegner" comes to mind here.[46] Cook-Lynn's refusal to read Wallace Stegner is not censure: it is agency to choose; it is a politics of refusal.

There is a sharp contrast between academic critique and a politics of refusal; they arise from different subject-positions, they affirm different values, and they operate through different relations. Eve Tuck and K. Wayne Yang, in their article "R-Words," define a politics of refusal this way: "Refusal is not just a 'no,' but a redirection to ideas otherwise unacknowledged or unquestioned."[47] This is the potential of a politics of refusal. It is not simply a "tear down" mechanism of discounting and dismissing, as academic critique may tend toward; instead, a politics of refusal dismantles in order to build.

Rather than asserting the machinations of normative critique, we might instead choose to take up that which disrupts. King offers an extended discussion of "postures, affective states (skepticism), and stubborn practices of insubordination such as refusal that frustrate forward movement and business-as-usual in the academy."[48] King calls us toward refusal and skepticism as "modes of engagement that are uncooperative and force an impasse in a discursive exchange."[49] Forcing an impasse prompts change. This is something other than discrediting a position; it is a form of interruption and redirection that creates momentum rather than immobility. A refusal to cooperate with the intellectual systems as they stand can guide our reading practice and our regard for text-making in the environmental humanities, as a primary pursuit.

The limits of knowing

Another white-settler motif that predominates literacy practices in the humanities is the presumption of entitlement to access. Tuck and Yang call this a "settler colonial configuration of knowledge that is petulantly exasperated by, and resentful of, limits." That decolonization limits knowledge should be seen as "productive, as indeed a good thing."[50] Here Tuck and Yang put into question what Linda Tuhiwai Smith, in *Decolonizing Methodologies*, calls "that most fundamental belief of all" within colonial orientations: "that individual researchers have an inherent right to knowledge and truth."[51]

A white researcher's sense of entitlement parallels the white reader's sense of entitlement. Melamed explicates this orientation in literary practices, stemming from the "institutionalizing of literature as a privileged apparatus for knowing difference."[52] Literature made it possible for "white liberal multiculturalism . . . to know and admire the cultural production of people of color" while failing to be accountable to people themselves.[53] These actual people become furthermore "replaced with racialized cultural products . . . and made illegible to the nation as the actual roots of grassroots democracy (in part) because they were not represented as such in dominant antiracist discourses."[54] Research and representation therefore become a stand-in not only for material change, but also for humanity itself. The onto-epistemologies and embodiments of human beings become invisibilized under their representation. The representation is taken up within liberal multiculturalist frameworks that are compatible with capitalist-colonial ideology, furthering the erasure.

Redressing this history requires a relational reading practice that is committed to the material. Our reading practices are ultimately about relationships, and thus they require an analysis of power. What is the relationship between the reader and what is being read? When a beneficiary of the dominant system reads a work of literature written by someone who is marginalized by that system, what is at stake? Is there deep respect for the work in its context, with full acknowledgment of continued histories and a desire to follow where the text leads, including if it leads to action and disruption in the world beyond the text?

Such questions invoke a reading practice that resists the appropriative tendencies of the humanities. Explicating appropriation in their introduction to the collection *Borrowed Power*, Bruce Ziff and Pratima V. Rao provide a set of guidelines: cultural appropriation is concerning when it harms the appropriated community or diminishes the vibrancy of a culture, when the impact of the appropriation damages or transforms a given cultural object or practice, when the appropriation wrongly allows some to benefit to the detriment of others, when stereotypes and controlling images are reinforced through the act of appropriation, when appropriation constitutes stealing the cultural forms of other for one's own prestige and profit, when the law fails to reflect alternative conceptions of rightful use.[55] In short, appropriation is problematic when the writer has a sense of an unchecked right to take, borrow, and steal without responsibility to the effects of their action—this sense stemming from privilege.

Gloria Anzaldúa, in the introduction to *Making Face, Making Soul*, demonstrates the power of writing that takes back what has been removed or pushed out. In so doing, she also draws a sharp distinction between appropriation born out of respect and appropriation that is in the service of dominant culture.

> In our *mestizaje* theories we create new categories for those of us left out or pushed out of the existing ones. We recover and examine non-Western aesthetics while critiquing Western aesthetics; recover and examine non-rational modes and "blanked-out" realities while critiquing rational, consensual reality; recover and examine indigenous languages while critiquing the "languages" of the dominant culture. And we simultaneously combat the tokenization and appropriation of our literatures and our writers/artists.[56]

Anzaldúa's words elucidate the difference between appropriation that co-opts and tokenizes and appropriation that recovers and examines what has been blanked-out, covered over, or subordinated. Anzaldúa continues, "The difference between appropriation and proliferation is that the first steals and harms; the second helps heal breaches of knowledge." She goes on to provide two examples. "the white [writer] writing about Native peoples or cultures displaces the Native writer and often appropriates the culture instead of proliferating information about it." Anzaldúa offers an

> example of a white woman who "rips off" people of color, examining Native spirituality and myth with a white collector's mentality. She passes off fiction as fact and distorts the true picture of native peoples in a way that they, as writers would not.[57]

Harms are enacted when the cultural significance of a story is not understood, or when the cultural practices determining rightful use are ignored: Lenore Keeshig-Tobias writes in the essay "Stop Stealing Native Stories":

> Stories are power. They reflect the deepest, the most intimate perceptions, relationships, and attitudes of a people. Stories show how a people, a culture, thinks. . . . So potent are stories that, in native culture, one storyteller cannot tell another's story without permission.

Warren Cariou's essay "Edgework: Indigenous Poetics as Re-Placement" echoes this point about the close relationship between story and identity: Quoting Louise Halfe's discussion of her book *Blue Marrow*, Cariou writes of "Cree-ing," which "is the pronunciation of identity, an assertion that the speaker is connected to her people, to a way of life that is larger than herself." Cariou continues, explaining how essential it is to understand story as tied to community: "I think each Indigenous nation expresses its identity and its uniqueness in a similarly active and verbal way. Poetry as a cry, a *cri*, that echoes through communities and through the land itself." While poetry may transcend cultural specificity, "moving across the lines of class and race and epistemology toward something more elemental in us all, something that we feel in our bodies like the sound of a drum,"[58] it also takes its power from the connection to culture, community, and land. Poetry and story are based in the community and place; to extract a story and appropriate it as one's own, without acknowledging its lineage and its ties to community, is to reduce that story and potentially do harm.[59]

How can we ensure that we do not fall into the traps and problems identified here? We can begin to ask important questions of ourselves, interrogating our motives as readers and writers. Who is the reading and writing for? Whom or what does the reading and writing serve? Have I done immersive research that has prompted me to come to terms with my assumptions, destabilizing the inherited ideas I have and master narratives I rely upon? In my research, have I moved beyond learning *about* to come to learn *from*, in a way that builds trust, without demand? Am I accountable?

Rosemary J. Coombe, in another chapter from the collection *Borrowed Power*, notes the profound problems with the common view that "everyone is implicitly equal in their capacity to write or be written about—to speak or be spoken for." Coombe demonstrates how

> Such a position purports to be apolitical, but manages only to be ahistorical and blind to relations of power. It ignores the very real social lines along which representation has been structured and the very real difficulties faced by certain social groups to represent themselves and speak on their own behalf.[60]

The appropriative practices of writing, reading, and research manifest what Nelson Maldonado-Torres calls the *ego conquiro* ("I conquer, therefore I am").[61] As Tuck and Yang explain, this term points to "the relationship of the conqueror's sense-of-self to his knowledge-of-others ('I know her, therefore I am me'). Knowledge of self-Others became the philosophical justification for the acquisition of bodies and territories, and the rule over them."[62] Tuck and Yang draw a line from this colonial trope to the activities of the academic–industrial complex:

The academy as a community of a practice that, *as a whole:*

1 Stockpiles examples of injustice, yet will not make explicit a commitment to social justice;
2 Produces knowledge shaped by the imperatives of the nation-state, while claiming neutrality and universality in knowledge production;
3 Accumulates intellectual and financial capital, while informants give a part of themselves away;
4 Absorbs or repudiates competing knowledge systems, while claiming limitless horizons.[63]

The reading and writing practices of the environmental humanities are complicit here. The narratives we seek and bring into each space we are part of, including classroom spaces, demand careful consideration: Do we prioritize pain narratives—narratives of environmental damage and destruction, narratives of fundamental loss—to the exclusion of narratives that represent multiple ways of being in resistance? Do we follow the desire of and leadership of the writers, scholars, and theorists who can tell us what role we might be able to play in support of the movements that are to remain theirs?

"Nothing about us without us" is therefore a core principle, and it pertains to choices made in the classroom. As Kathie Irwin (Māori) is quoted saying in Linda Tuhiwai Smith's *Decolonizing Methodologies:*

> We don't need anyone else developing the tools which will help us to come to terms with who we are. We can and will do this work. Real power lies with those who design the tools—it always has. This power is ours.[64]

We need classroom spaces that are oriented toward this fact. This is in sharp contrast to the current structures of education, which erect barriers, co-opt, and centralize power in institutional interests. Where the classroom should be a space of vibrant activity, educational spaces, established as arms of colonialism, are characterized by surveillance, acculturation, and assimilation in a disciplinary system. As Tuhiwai Smith notes,

> The dehumanizing tendencies within imperial and colonial practices are deeply encoded. These practices serve constantly to deny that colonized people actually have ideas of our own, can create new ideas, and have a rich knowledge base from which to draw.[65]

The colonial logic "that it is simply impossible, ridiculous even, to suggest that the object of research can contribute anything"[66] also plays out in the classroom, where the "object of learning" (i.e., the student) is positioned similarly. As bell hooks elucidates, "In our nation most colleges and universities are organized around the principles of dominant culture. . . . Concurrently, students are encouraged to doubt themselves, their capacity to know, to think, and to act"[67]; rather than locating knowledge-practices within themselves and in their home communities, they are enculturated to locate knowledge in the mind of those who are authorized by the institution—a predominantly white faculty. This fundamental racism plays out across the overlapping spheres of academic work: classroom practices, research and scholarship, administrative decision-making, and so on. Assumptions regarding who the knowledge-holders are, who can be entrusted to lead, whose perspectives matter in decision-making practices, where resources should go, and so on—all reinstitute this intersectional racism day after day. We need to remake the university in order that, in Tuhiwai Smith's words, the "[c]ommunity itself makes its own definitions."[68] This means that white people need to daily leverage or give up their seat at the table; doing this not as a performance of individual virtue, but to keep the focus on systemic change and righting the wrongs that have created forms of underrepresentation.

On the idea of a "decentered" classroom

Critical pedagogy, following the work of Paolo Freire, popularized the idea of a decentered classroom. As the term "decentered" circulates in white institutional spaces, it carries with it naturalizing assumptions: about who the person in the front of the room is and that they have a level of privilege that can accommodate stepping to the side, stepping out of the center. This framework allows for a belief in the "consciousness raising" work of the white faculty member who wears the pedantry of "knowing best" in the dress clothes of progressive ideals.

To truly decenter would be to go further and destabilize white epistemology as it plays out in classroom policies, grading and assessment, language practices, learning content, the idea of "assigning" work, and so on, but this depth-work is rarely discussed and grappled with. In what bell hooks calls the "conventional dominator classroom,"[69] a single privileged knower (the teacher) makes the rules and dictates what takes place in the classroom—setting the syllabus, planning the learning, without consultation and dialogue with the students who make up the majority of the classroom community.

True decentering would entirely re-position the white educator and all they feel entitled to contribute to a classroom space. Who am I to make decisions about what the people in this room need to read and write? Who am I to know what the people in this room need for the journey they are on? In *No More Heroes*, Jordan Flaherty recounts an interview with Caitlin Breedlove, a white antiracist organizer who puts it clearly: "The assumption that you have something to contribute to communities you know nothing about is 'an incredibly entitled notion.'"[70]

As multiple students of color have reminded me, getting out of the way may be what is required of me as a white educator. "Move out the way, listen, learn," a former student said to me recently, reflecting on what she most wanted from her white faculty. To honor her words, I want to recognize that moving aside in order to respectfully create and hold space is an act made in acknowledgment that I never should have been centered in the first place. The white teacher has work to do: to recognize the operations of power moment-by-moment; to leverage privilege in ways that are transformative of inequitable distribution of benefits; and to support ongoing movement-building. Breedlove explains the role white people can have in transformative movements: "What we need are strategies that don't make us look as good, don't make us visible."[71] To carry forward Breedlove's recommendation requires a persistent vigilance against noblesse oblige orientations, the self-flagellating glory of white interior transformation, the pursuit of recognition (being crowned an "ally"), overestimated contributions, or presumptive forms of "help."

One of my mentors once told me: we need classrooms that are drenched in a notion of freedom—or the closest thing we can get to this within the space of the classroom. What would a pedagogy look like that focused on envisioning a world in which *I am free and you are free*?[72] These words echo bell hooks whose brilliant trilogy of pedagogy books offer a model of "education as the practice of freedom." hooks critiques schooling as a site that actively supports imperialist white-supremacist capitalist patriarchy.[73] The dominator classroom often relies on what we might call a "monarch model": "To many professors of all races," she writes, "the classroom is viewed as a mini-country governed by their autocratic rule,"[74] with predetermined outcomes under the paternalistic assumption that one mind can know what is in the best interest of any group of people.

Theorists including bell hooks and Bettina Love identify the ways in which these pedagogies are aligned with systems of punishment. "Subordinated by a hierarchical system that indoctrinates students early on," hooks writes, students "know that their success depends on their capacity to obey."[75] In their abolitionist manifesto on the university, Boggs et al. suggest this same comparison in the form of a question: "Are prisons and universities two sides of the same coin?" They reflect: "When we raise this question, does it make you anxious? We feel this anxiety, too, and we want to sit with it, to grapple with the impasse such questions open up."[76] To overturn this long-held function of education requires dismantling it brick by brick.

There are many complexities here. The hallmark of a decentered classroom is greater levels of classroom engagement, with the accompanying belief that learning resides with the temporary community brought together in the classroom space. But community is hard-won, difficult to build. "Community," Mohanty notes,

> is the product of work, of struggle; it is inherently unstable, contextual; it has to be constantly reevaluated in relation to critical political priorities; and it is the product of interpretation, interpretation based on an attention to history.[77]

A community needs to work through the vexed matters of finding trust and recognition and a shared language for addressing harms. A skillful facilitator can create the conditions where a temporary community might be forged, but this is difficult work—and this is just one line in the multidimensional efforts of building a learning experience.

There is much to account for: some of the students who come to the classroom will arrive in full allegiance to dominant orientations, and what they say and do in the space may create harms. The instructor may decide that lecturing or creating few opportunities for student-to-student interactions is the way to mitigate these circumstances. We want to do what we can to approximate safety in the classroom, but we know that the university environment is not safe. We're grappling with safety and transformation in the first-, second-, and third-person position—meaning, as Patricia Owen-Smith explains: first-person referring to personal experiences, second-person referring to collaborative dialogue, and third-person referring to the material engaged from outside.[78] There is much at issue in each classroom space, and this reality calls us to go slow, to think through each choice, to find support where we can, and to stay present to the complexities of facilitation as an embodied practice.

The work is to make together what an antiracist practice can look like, to envision a radically different model that pushes against institutional conventions and breaks out of the imperial-institutional will to control. While it is beyond the scope of this text to fully imagine what this looks like (although the next chapter takes up some of these questions)—and in my own classes I only approximate this vision—there are many books that we can stay near as we do the work of remaking: books on education by Bettina Love, bell hooks, Eli Meyerhoff, and more.[79] Indeed, these books may be those we wish to bring into the educational space and share with students in order to think collectively about what we come to the classroom to do, to understand the potential of the space.

The texts that come before us

The texts we bring into the classroom have great shaping force. Curated reading lists that prioritize the work of queer and trans writers, Black and Indigenous writers, people of color, and Global South writers are powerful in opening possibilities. We can bring these lists to the shared classroom table (if only our classrooms could exist as a single, accessible table—rather than forcefully individualized desks which find their parallel in the Zoom virtual conferencing platform's virtual squares) for collaboration, remaking, and building. Flexibility in course format options can allow for ad hoc reading circles around particular texts of interest; these make space to collectively and responsively decide where discussions need to go next—what questions one group or another wishes to ask and what texts might support those questions.

In building reading lists, which serve as only a starting point, we can ask what world does this text invite students to inhabit? On whose terms does it operate? The work is to select texts that forward a decolonizing understanding of the world from multiple cultural traditions, that bring our attention to epistemic difference: for example, reading the essays collected in Lara Medina and Martha R. Gonzales' *Voices from the Ancestors: Xicanx and Latinx Spiritual Expressions and Healing Practices* (2019); Laura Harjo's *Spiral to the Stars: Mvskoke Tools of Futurity* (2019); Angayuqaq Oscar Kawagley's *A Yupiaq Worldview: A Pathway to Ecology and Spirit* (2006); and Na'im Akbar's *Know Thy Self* (1999).[80] Reading through a list like this produces a sense of many-worldedness within a curriculum. These texts bring us to face the cultural contingency of knowing, as they also refuse to adapt to or accommodate dominant epistemologies. In approaching texts, we need cognizance of the cosmologies, epistemologies, and ontologies that the text is enmeshed with and potentially revitalizing and troubling. *Awareness of* here is different from a consuming or essentializing *knowledge of*. "Awareness of" is a present, unfolding attention to what the text resists, to its complexity, and to the ways that it cannot be fully known. At the same time, self-reflection is necessary as is a systemic analysis of the academic situation: how dominant epistemologies may lie in wait, ready to take up the text through evaluative, homogenizing, appropriating, or dismissive means. Epistemologies cannot be categorized or summarized, but a reader can begin to feel the full dimensionality of them by encountering texts that do not accommodate dominant knowledge regimes. This kind of encounter does not resolve into knowledge but instead remains in the practice articulated by the Zapatistas and identified earlier: caminando preguntamos; *walking we ask questions*. We move through a text; we read; we participate in movement-building; we stay with an openness to all we do not know; and we continue to move and to ask questions. This orientation subverts a mastery paradigm while precluding immobilization in the presence of the gravity and complexity of what is before us.

As we engage in the reception and production of texts among us, we need to guard against some of the common tendencies in colonial knowledge practices. Dorothy Wang in the book *Thinking Its Presence* names several of these tendencies:

treating the writing of BIPOC authors as a transparent window into an ethnographic "truth,"[81] rendering these authors as "abstract, one-dimensional, homogenous or interchangeable,"[82] "ignor[ing] the ethnicity of the poet, even when the poet makes clear that racialized/ethnic identity is not a trivial concern in the work,"[83] "[c]onjuring up a link that is not there (or artificially 'imposing' the issue of race . . . usually by smuggling in some reductive essentialist version of racial identity),"[84] or ignoring racism as an embarrassment.

As an antidote to what she describes earlier, Wang encourages us toward "praxis-based critical argumentation, in which the poems [or texts] themselves suggest theoretical orientations, resist abstract generalization that can easily oversimplify and render reductive and one-dimensional arguments about racial subjectivity and minority poetry."[85]

Wang's ideas here echo a central idea in McKittrick's *Dear Science:* to "honor the creative text as a theoretical text."[86] With this regard, the text, if it invites, might break open the regimes of thinking that it cannot accommodate. Wang and McKittrick lead us toward reading practices that honor the text and understand the epistemologies embedded within it, while not seeking to consume the text or subsume it into one's existing sense of self-in-world. We should read with the humility that acknowledges difference and seeks to orient toward what is offered without a demand for knowability. This humility follows where the text leads, but does not demand that the text be incorporable; it respects the integrity of the text and its opacity as it is embedded in epistemologies that far exceed it.

We can approach each interlocutor and each idea with cognizance of all we do not know about its history and relationality. Indeed, we might approach it the way Robin Wall Kimmerer describes approaching a plant in the Honorable Harvest:

> When we encounter the plant . . . we introduce ourselves and explain our purpose. . . . If one asks permission, then one must also listen for the answer. This kind of listening may involve many different kinds of perception.[87]

This entails respect without a predetermined answer. How does the text give permission? What does it invite? How might it open spaces for "thinking otherwise"? What does it mean to cultivate attentiveness to the experience of engaging in reading–writing–thinking and how new encounters might introduce new forms into the mind? The work of reading–writing–thinking is, after all, a way to "share our lessons of unknowing ourselves," McKittrick notes, speaking to a "we" of Black authors and scholars. In this, we "refuse what they want us to be; we risk reading what we cannot bear and what we love too much and then we let it go, revise, and read it again."[88]

The forms of reading and writing described here are deeply participatory. They call one's attention to inner landscapes. They are embodied, allowing for the full self to come into the encounter. They invite the writer–reader to find their way through multiple ways of knowing and ways of being. They are present to text and text-making, able to suspend what is predetermined in order to find what it means to come-to-know. These practices are conscious of positionality while at the same time being not driven by the ego. We might say that these practices are mindful.

Our coursework in the environmental humanities should be a resource for living: The knowledge is in service of the people who come to it, and not the other way around; that is, the students are not subordinated to the education; education is to serve them in their ongoing work.

The motley nature of a reading list enacts a practice that is akin to what Ana Louise Keating describes in *Transformation Now:* Keating's book brings into conversation authors who represent a range of identities, times, and places. She finds connections between different thinkers, drawing "disparate worldviews into dialogue" and bringing together "areas of thought and perspectives that are too often kept far apart."[89] Keating's practice is "[g]rounded in a framework of interconnectivity," she writes, linking this practice to that of Anzaldúa's "nepantlera, torn between theories, divided among worldviews, in love with authors and philosophers who rarely (if ever) speak to one another or even acknowledge one another's work."[90]

Keating's mode, framed differently, might be understood as a sibling practice to what VèVé Clark calls "diaspora literacy": "the ability to read and comprehend multifarious and overlapping discourses of the Black diaspora from an informed and learned perspective."[91] This practice of reading for "complex commonality so spacious that it embraces difference—even apparently mutually exclusive differences"[92] is done with full regard for the risks of an "add and stir" (to use Chandra Mohanty's phrase) melting-pot multiculturalism. The practice of reading determinedly preserves difference—specifically, "difference that makes a difference"[93]—that is, the difference that destabilizes enforced regimes of knowing.

I read Keating's practice as a response to the bind that is usefully identified by Chandra Mohanty who writes:

> how do we undermine the notions of multiculturalism as melting pot, or multiculturalism as cultural relativism that so permeate U.S. consumer culture and that are mobilized by the corporate academic as a form of containment, and practice a multiculturalism that is about the decolonization of received knowledges, histories, and identities, a multiculturalism that foregrounds questions of social justice and material interests, which actively combats the hegemony of global capital.[94]

Moving beyond liberal multiculturalism means enacting onto-epistemic justice in the classroom in ways that are trained on material change. It means honoring the methodologies and embodiments that are often unacknowledged and erased in the classroom. It means dwelling on the ways in which environmental issues are understood through different frameworks, by different stakeholders, in ways that prioritize the worldsenses of Indigenous people, peoples of the Afro-Caribbean diaspora, peoples of the Global South. It should be clear that the developmentalist viewpoints of the NGO or the multinational conglomerate are not on an equal plane with Indigenous worldviews these enterprises have historically perpetuated violence against. An "all views are equal" and "all of us are equal" relativist approach only serves to cover over the gross disparities that structure our world, while keeping these structures in place. The viewpoints of shared texts in a classroom should be considered against this rubric.

We need voices that dwell in possibility and transformative potential. McKittrick's intention is to bring into her classroom that which has the quality of felt possibility: "I feel possibility. I think this is what possibility feels like," she writes

> It is not narrated textually. It is not an idea. I have always pushed myself and my students to think about how black liberation can be understood as a site of possibility and as a struggle that is ongoing but never resolved.[95]

The texts we invite into the space can help us attend to multiple dimensions of the causes, consequences, and possible future resolutions of environmental justice struggles. They can move us toward transformation by drawing on multiple traditions that respect multiple ontologies and epistemologies. We can read and write in a way that acknowledges how language diversity, cultural diversity, and biodiversity are related. We can come to understand how writers relate differently to the places they write with and within, based on history. We can learn to question and speak back to dominant narratives. We can see how our understanding of the world is conditioned by the identities, contexts, experiences, and perspectives we inhabit. We can grapple with what is at stake in our knowing as we imagine and encounter possibilities that transcend that knowing.

Notes

1 Boaventura de Sousa Santos, *The End of the Cognitive Empire: The Coming of Age of Epistemologies of the South* (Duke UP, 2018).
2 Jodi Melamed, *Represent and Destroy: Rationalizing Violence in the New Racial Capitalism* (U of Minnesota P, 2011) 105, 49.
3 Melamed, *Represent and Destroy*, 49.
4 Erin James, *The Storyworld Accord: Econarratology and Postcolonial Narratives* (U of Nebraska P, 2015) xi.
5 James, *Storyworld Accord*, xi.
6 Melamed, *Represent and Destroy*, 107. Linda Tuhiwai Smith makes a similar point, writing that: "Imperialism cannot be struggled over only at the level of text and literature. Imperialism still hurts, still destroys and is reforming itself constantly." Linda Tuhiwai Smith, *Decolonizing Methodologies*, 2nd edition (Zed Books, 2012) 20.
7 Melamed, *Represent and Destroy*, 16. The national attention on the canon wars is evidence of how the potential of literary study has been understood. Jodi Melamed argues that the war—commonly narrated as a battle between the conservative Great Books versus a multicultural and diverse orientation—is actually a battle between two fundamentally conservative strategies: "with canon defenders advocating for an assimilative pluralism and canon integrationists for a positive pluralism" (34)—positive pluralism being that which "proposed reading literature to be an effective strategy to recognize and unify the multiple cultures of the U.S. nation" (112). Melamed continues: "Whereas Great Books advocates disfigured the epistemo-intellectual work of racial antiracist social movements through

attack and ridicule, the liberal multiculturalism obscured such practices by supporting and misrepresenting them . . . miscasting it as merely ethnic self-expression" (113) with the requisite "respect for multiple identities conceived as cultural property" (108). From this synopsis of Melamed's argument the stakes of reading practices become clear.

8 Melamed, *Represent and Destroy*, 3.
9 Melamed, *Represent and Destroy*, 36.
10 Melamed, *Represent and Destroy*, 80.
11 Melamed, *Represent and Destroy*, 36.
12 Melamed, *Represent and Destroy*, 229.
13 Katherine McKittrick, *Dear Science and Other Stories* (Duke UP, 2020) 8.
14 McKittrick, *Dear Science*, 16.
15 McKittrick, *Dear Science*, 7.
16 Claudia Rankine and Beth Loffreda, *The Racial Imaginary: Writers on Race and the Life of the Mind* (Fence Books, 2015) 14.
17 Rankine and Loffreda, *The Racial Imaginary*, 22.
18 Rauna Kuokkanen, *Reshaping the University: Responsibility, Indigenous Epistemes, and the Logic of the Gift* (U of British Columbia P, 2007) 4.
19 Rankine and Loffreda, *The Racial Imaginary*, 22.
20 Rankine and Loffreda, *The Racial Imaginary*, 16. Similarly, Zenju Earthlyn Manuel identifies how this tendency to enforce the fantasy of transcendence shows up in white discourses of spirituality as well.
21 Chandra Talpade Mohanty, *Feminism Without Borders: Decolonizing Theory, Practicing Solidarity* (Duke UP, 2003) 102.
22 Dorothy Wang explicates a maintained binary in literary criticism that divides the transcendent aesthetic text and the political text, which, Dorothy Wang notes, is a racist division that keeps white texts centered as "true literature" and invisibilizes the politics of the text labeled as aesthetic. Wang articulates the "literary versus cultural" divide for its investments in hierarchizing texts and their knowers, in turn obscuring white politics.
23 M. NourbeSe Philip, in the essay "The Disappearing Debate," explains how ideas about a free imagination are tied to privilege: "To state the obvious, in a racist, sexist and classist society, the imagination, if left unexamined, can and does serve the ruling ideas of the time."
24 Rankine and Loffreda, *The Racial Imaginary*, 20.
25 Rankine and Loffreda, *The Racial Imaginary*, 18.
26 In *Emergent Strategy*, adrienne maree brown says this about the pervasive culture of critique: "Staying focused on our foundational miraculous nature is actually very hard work in our modern culture of deconstruction. We are socialized to see what is wrong, missing, off, to tear down the ideas of others and uplift our own." She encourages critique only from the position of participation: "Critique as a *participant* who is shaping the work. Be willing to do whatever task is required of you, whatever you are capable of." See adrienne maree brown, *Emergent Strategy: Shaping Change, Changing Worlds* (AK Press, 2017) 5, 119–120.
27 Rita Felski, *The Limits of Critique* (U of Chicago P, 2015) 3.
28 Felski, *Limits of Critique*, 5.
29 Felski, *Limits of Critique*, 13.
30 Felski, *Limits of Critique*, 143, 147.
31 Jessica Lopez Lyman, "Imposter Poet: Recovering from Graduate School," *How Dare We! Write: A Multicultural Creative Writing Discourse*, edited by Sherry Quan Lee (Modern History, 2017) 17.
32 The critical stance can dismiss and negate anything in its path: Kuokkanen notes the white tendency to call equity work "anti-intellectual dogma unworthy of exploration or discussion" (McIntyre 193 quoted in Kuokkanen, *Reshaping the University*, 18). See Sheila McIntyre, "Studied Ignorance and Privileged Innocence: Keeping Equity Academic," *Canadian Journal of Women and the Law* 12.1 (2000): 193.

33 Kandace Creel Falcón, "What Would Edén Say? Reclaiming the Personal and Grounding Story in Chicana Feminist (Academic) Writing," *How Dare We! Write: A Multicultural Creative Writing Discourse*, edited by Sherry Quan Lee (Modern History, 2017) 10.
34 Paul Ricœur, *The Symbolism of Evil*, translated by Emerson Buchana (Beacon, 1969).
35 Tiffany Lethabo King, "Humans Involved: Lurking in the Lines of Posthumanist Flight," *Critical Ethnic Studies* 3.1 (Spring 2017): 182.
36 Abigail Boggs, et al., *Abolitionist University Studies: An Invitation* (2019), https://abolition.university/invitation/ 11.
37 Boggs, et al., *Abolitionist University*, 27.
38 Tema Okun, "White Supremacy Culture," *DRWorksBook* (2019), www.dismantlingracism.org/white-supremacy-culture.html.
39 King, "Humans Involved," 174.
40 Chris Finley, "Building Maroon Intellectual Communities," *Otherwise Worlds: Against Settler Colonialism and Anti-Blackness*, edited by Tiffany Lethabo King, Jenell Navarro, and Andrea Smith (Duke UP, 2020) 364.
41 King, "Humans Involved," 166.
42 King, "Humans Involved," 165.
43 King, "Humans Involved," 170.
44 Eve Tuck and K. Wayne Yang, "R-Words: Refusing Research," *Humanizing Research: Decolonizing Qualitative Inquiry with Youth and Communities*, edited by Django Paris and Maisha T. Winn (Sage, 2014) 224.
45 King, "Humans Involved," 172.
46 Elizabeth Cook-Lynn, *Why I Can't Read Wallace Stegner and Other Essays: A Tribal Voice* (U of Wisconsin P, 1996).
47 Tuck and Yang, "R-Words," 239.
48 King, "Humans Involved," 163.
49 King, "Humans Involved," 164.
50 Tuck and Yang, "R-Words," 239.
51 Linda Tuhiwai Smith, *Decolonizing Methodologies*, 2nd edition (Zed Books, 2012) 176.
52 Melamed, *Represent and Destroy*, xvi, 15.
53 Melamed, *Represent and Destroy*, 116.
54 Melamed, *Represent and Destroy*, 39.
55 Bruce H. Ziff and Pratima V. Rao, *Borrowed Power: Essays on Cultural Appropriation* (Rutgers UP, 1997).
56 Gloria Anzaldúa, *Making Face, Making Soul: Haciendo Caras: Creative and Critical Perspectives by Feminists of Color* (Aunt Lute, 1990) xxvi.
57 Anzaldúa, *Making Face*, xxi.
58 Warren Cariou, "Edgework: Indigenous Poetics as Re-Placement," *Indigenous Poetics in Canada*, edited by Neal McLeod (Wilfrid Laurier UP, 2014) 32. Louise Halfe, *Blue Marrow* (Couteau, 2004).
59 When a piece is divorced from its lineage and its tie to community, there is the risk of misrepresenting. Lenore Keeshig-Tobias warns the writer who wants to adopt the stories of Indigenous peoples, stories that are not the writer's own: "If you want to write our stories, then be prepared to live with us. And not just for a few months."
60 Rosemary J. Coombe, "The Properties of Culture and the Possession of Identity: Postcolonial Struggle and the Legal Imagination," *Borrowed Power: Essays on Cultural Appropriation*, edited by Bruce Ziff and Pratima V. Rao (Rutgers, 1997) 78.
61 Nelson Maldonado Torres, *Against War: Views from the Underside of Modernity* (Duke UP, 2008). Quoted in Tuck and Yang, "R-Words," 224.
62 Tuck and Yang, "R-Words," 224. Echoing this point and quoting Oswald de Andrade, Doris Sommer puts it this way: "I am only interested in what does not belong to me. Law of Man. Law of the Cannibals." This tendency Sommer attributes to a colonial mentality: "Appropriation of the other is what our New World cultures feed on." Sommer continues, speaking of an "us" of colonial moderns: "devouring alterity is what makes us modern (or just human, for Freud), as we participate in an Occidental culture nourished

on novelty." Doris Sommer, *Proceed with Caution, When Engaged by Minority Writing in the Americas* (Harvard UP, 1999) 11.
63 Tuck and Yang, "R-Words," 233.
64 Quoted in Smith, *Decolonizing Methodologies*, 40.
65 Smith, *Decolonizing Methodologies*, 203.
66 Smith, *Decolonizing Methodologies*, 64.
67 bell hooks, *Teaching Community: A Pedagogy of Hope* (Routledge, 2003) 130.
68 Smith, *Decolonizing Methodologies*, 129.
69 hooks, *Teaching Community*, 8.
70 Jordan Flaherty, *No More Heroes: Grassroots Challenges to the Savior Mentality* (AK Press, 2016) 11. Along similar lines in the same text, Flaherty interviews Paulina Helm-Hernandez, a queer Latina organizer and artist who says this: "I think white people in movement building need to make a call about whether they will be individual activists or if they are really ready to commit to collective organizing. The latter means that you don't have to always be the final vote on the strategy, pace, timing, tone, and approach. Put another way, it means you have to learn how to share political imagination, power, and work without always having to be in charge." Quoted in Flaherty, *No More Heroes*, 12.
71 Flaherty, *No More Heroes*, 14.
72 Thanks to Marina Smerling.
73 hooks, *Teaching Community*, xiii.
74 hooks, *Teaching Community*, 85.
75 hooks, *Teaching Community*, 86.
76 Boggs, et al., *Abolitionist University*, 2.
77 Chandra Talpade Mohanty, *Feminism Without Borders: Decolonizing Theory, Practicing Solidarity* (Duke UP, 2003) 104.
78 Patricia Owen-Smith, *The Contemplative Mind in the Scholarship of Teaching and Learning* (Indiana UP, 2017) 9.
79 See bell hooks, *Teaching to Transgress: Education as the Practice of Freedom* (Routledge, 1994); Bettina L. Love, *We Want to Do More than Survive: Abolitionist Teaching and the Pursuit of Educational Freedom* (Beacon, 2019). See also Eli Meyerhoff, *Beyond Education: Radical Studying for Another World* (U of Minnesota P, 2019) and Robert H. Haworth and John M. Elmore, *Out of the Ruins: The Emergence of Radical Informal Learning Spaces* (PM Press, 2017). Thanks to Renée Byrd for some of these references.
80 Thanks to John Johnson for the reference to Na'im Akbar, *Know Thy Self* (Mind Productions, 1998).
81 Dorothy J. Wang, *Thinking Its Presence: Form, Race, and Subjectivity in Contemporary Asian American Poetry* (Stanford UP, 2014) 23.
82 Wang, *Thinking Its Presence*, 37.
83 Wang, *Thinking Its Presence*, 22.
84 Wang, *Thinking Its Presence*, 37.
85 Wang, *Thinking Its Presence*, 39.
86 McKittrick, *Dear Science*, 52.
87 Robin Wall Kimmerer, "*Mishkos Kenomagwen,* the Lessons of Grass: Restoring Reciprocity with the Good Green Earth," *Traditional Ecological Knowledge: Learning from Indigenous Practices for Environmental Sustainability*, edited by Melissa K. Nelson and Dan Shilling (Cambridge UP, 2018) 33.
88 McKittrick, *Dear Science*, 34.
89 AnaLouise Keating, *Transformation Now! Toward a Post-Oppositional Politics of Change* (U of Illinois P, 2013) 18.
90 Keating, *Transformation Now*, 18.
91 VèVè Clark, "Developing Diaspora Literacy and Marasa Consciousness," *Comparative American Identities: Race, Sex, and Nationality in the Modem Text*, edited by Hortense J. Spillers (New York: Routledge, 1991), 40–61. Summarized in McKittrick, *Dear Science*, 29.
92 Keating, *Transformation Now*, 11.

93 Melamed, *Represent and Destroy*, 229.
94 Mohanty, *Feminism without Borders*, 188.
95 McKittrick, *Dear Science*, 71.

Works cited

Akbar, Na'im. *Know Thy Self.* Mind Productions, 1998.

Alcoff, Linda Martín. "Epistemologies of Ignorance: Three Types." *Race and Epistemologies of Ignorance*, edited by Shannon Sullivan and Nancy Tuana. SUNY P, 2007.

Anzaldúa, Gloria. *Making Face, Making Soul: Haciendo Caras: Creative and Critical Perspectives by Feminists of Color.* Aunt Lute, 1990.

Boggs, Abigail, Eli Meyerhoff, Nick Mitchell, and Zach Schwartz-Weinstein. *Abolitionist University Studies: An Invitation*, 2019. https://abolition.university/invitation/

Cariou, Warren. "Edgework: Indigenous Poetics as Re-Placement." *Indigenous Poetics in Canada*, edited by Neal McLeod. Wilfrid Laurier UP, 2014, pp. 31–38.

Clark, VèVè. "Developing Diaspora Literacy and Marasa Consciousness." *Comparative American Identities: Race, Sex, and Nationality in the Modem Text*, edited by Hortense J. Spillers. Routledge, 1991, pp. 40–61.

Cook-Lynn, Elizabeth. *Why I Can't Read Wallace Stegner and Other Essays: A Tribal Voice.* U of Wisconsin P, 1996.

Coombe, Rosemary J. "The Properties of Culture and the Possession of Identity: Postcolonial Struggle and the Legal Imagination." *Borrowed Power: Essays on Cultural Appropriation*, edited by Bruce Ziff and Pratima V. Rao. Rutgers, 1997, pp. 74–97.

Falcón, Kandace Creel. "What Would Edén Say? Reclaiming the Personal and Grounding Story in Chicana Feminist (Academic) Writing." *How Dare We! Write: A Multicultural Creative Writing Discourse*, edited by Sherry Quan Lee. Modern History, 2017.

Felski, Rita. *The Limits of Critique.* U of Chicago P, 2015.

Finley, Chris. "Building Maroon Intellectual Communities." *Otherwise Worlds: Against Settler Colonialism and Anti-Blackness*, edited by Tiffany Lethabo King, Jenell Navarro, and Andrea Smith. Duke UP, 2020, pp. 362–369.

Flaherty, Jordan. *No More Heroes: Grassroots Challenges to the Savior Mentality.* AK Press, 2016.

Freire, Paolo. *Pedagogy of the Oppressed.* 30th Anniversary ed. Continuum, 2000.

Halfe, Louise. *Blue Marrow.* Couteau, 2004.

Harjo, Laura. *Spiral to the Stars: Mvskoke Tools of Futurity.* U of Arizona P, 2019.

Haworth, Robert H., and John M. Elmore. *Out of the Ruins: The Emergence of Radical Informal Learning Spaces.* PM Press, 2017.

hooks, bell. *Teaching to Transgress: Education as the Practice of Freedom.* Routledge, 1994.

hooks, bell. *Teaching Community: A Pedagogy of Hope.* Routledge, 2003.

James, Erin. *The Storyworld Accord: Econarratology and Postcolonial Narratives.* U of Nebraska P, 2015.

Kawagley, Angayuqaq Oscar. *A Yupiaq Worldview: A Pathway to Ecology and Spirit.* 2nd ed. Waveland P, 2006.

Keating, AnaLouise. *Transformation Now!: Toward a Post-Oppositional Politics of Change.* U of Illinois P, 2013.

Keeshig-Tobias, Lenore. "Stop Stealing Native Stories." *Borrowed Power: Essays on Cultural Appropriation*, edited by Bruce Ziff and Pratima V. Rao. Rutgers, 1997, pp. 71–74.

Kimmerer, Robin Wall. "*Mishkos Kenomagwen*, the Lessons of Grass: Restoring Reciprocity with the Good Green Earth." *Traditional Ecological Knowledge: Learning from Indigenous Practices for Environmental Sustainability*, edited by Melissa K. Nelson and Dan Shilling. Cambridge UP, 2018, pp. 27–57.

King, Tiffany Lethabo. "Humans Involved: Lurking in the Lines of Posthumanist Flight." *Critical Ethnic Studies* 3.1 (Spring 2017): 162–185.

Kuokkanen, Rauna. *Reshaping the University: Responsibility, Indigenous Epistemes, and the Logic of the Gift*. U of British Columbia P, 2007.

Love, Bettina L. *We Want to Do More Than Survive: Abolitionist Teaching and the Pursuit of Educational Freedom*. Beacon, 2019.

Lyman, Jessica Lopez. "Imposter Poet: Recovering from Graduate School." *How Dare We! Write: A Multicultural Creative Writing Discourse*, edited by Sherry Quan Lee. Modern History, 2017.

Manuel, Zenju Earthlyn. *The Way of Tenderness: Awakening through Race, Sexuality, and Gender*. Wisdom, 2015.

McIntyre, Sheila. "Studied Ignorance and Privileged Innocence: Keeping Equity Academic." *Canadian Journal of Women and the Law* 12.1 (2000).

McKittrick, Katherine. *Dear Science and Other Stories*. Duke UP, 2020.

Medina, Lara, and Martha R. Gonzales. *Voices from the Ancestors: Xicanx and Latinx Spiritual Expressions and Healing Practices*. U of Arizona P, 2019.

Melamed, Jodi. *Represent and Destroy: Rationalizing Violence in the New Racial Capitalism*. U of Minnesota P, 2011.

Meyerhoff, Eli. *Beyond Education: Radical Studying for Another World*. U of Minnesota P, 2019.

Mignolo, Walter D. "Epistemic Disobedience, Independent Thought and Decolonial Freedom." *Theory, Culture, and Society* 26.7–8 (2009): 159–181.

Mills, Charles W. *The Racial Contract*. Cornell UP, 1997.

Mohanty, Chandra Talpade. *Feminism Without Borders: Decolonizing Theory, Practicing Solidarity*. Duke UP, 2003.

Okun, Tema. "White Supremacy Culture." *DRWorksBook*, 2019. www.dismantlingracism.org/white-supremacy-culture.html

Owen-Smith, Patricia. *The Contemplative Mind in the Scholarship of Teaching and Learning*. Indiana UP, 2017.

Philip, M. NourbeSe. "The Disappearing Debate: Or How the Discussion of Racism Has Been Taken Over by the Censorship Issue." *Borrowed Power: Essays on Cultural Appropriation*, edited by Bruce Ziff and Pratima V. Rao. Rutgers, 1997, pp. 97–109.

Rankine, Claudia, and Beth Loffreda. *The Racial Imaginary: Writers on Race and the Life of the Mind*. Fence, 2015.

Ricœur, Paul. *The Symbolism of Evil*, translated by Emerson Buchana. Beacon, 1969.

Santos, Boaventura de Sousa. *The End of the Cognitive Empire: The Coming of Age of Epistemologies of the South*. Duke UP, 2018.

Smith, Linda Tuhiwai. *Decolonizing Methodologies*. 2nd ed. Zed Books, 2012.

Sommer, Doris. *Proceed with Caution, When Engaged by Minority Writing in the Americas*. Harvard UP, 1999.

Torres, Nelson Maldonado. *Against War: Views from the Underside of Modernity*. Duke UP, 2008.

Tuck, Eve, and K. Wayne Yang. "R-Words: Refusing Research." *Humanizing Research: Decolonizing Qualitative Inquiry with Youth and Communities*. Sage, 2014, pp. 223–248.

Wang, Dorothy J. *Thinking Its Presence: Form, Race, and Subjectivity in Contemporary Asian American Poetry*. Stanford UP, 2014.

Ziff, Bruce H., and Pratima V. Rao. *Borrowed Power: Essays on Cultural Appropriation*. Rutgers UP, 1997.

5
GATHERING TOGETHER

Higher education can be a lever for changing the knowledge regimes that deny ethos and that negate and destroy eco-social worlds. To take this orientation is to trouble the university, confronting its conventions. How educational spaces are configured, who is invited to them, who is positioned as credible within them, are some of the most necessary questions. Can our classrooms become spaces for coalition-building and for grappling with the complexities of coexistence in a pluriversal world?

Acknowledging our position as those paid and trained by an epistemically violent system that is tailored for its own self-preservation, what capacity do we have for transformative work? Many of us, given our disciplinary training, have been taught to be the keepers of white supremacist knowledges, passed down through the "rigor" (a coded term) of doctoral programs that tightly suture "expertise" to a canonized set of thinkers. The repetition of academe's conventions is one way the university's history is carried forward. The workings of knowledge/power cannot be simply sloughed off. True disruption of this reproduction does not go unnoticed. Interruption means an end to business-as-usual models in this business of education. It also means reconfiguring how some of us understand ourselves in educational spaces. Rauna Kuokkanen recounts a conversation where a non-Indigenous faculty member identified the following:

> In my experience, when indigenous perspectives are genuinely included in the curriculum and the classroom, the epistemic and pedagogical changes involved are huge. I believe that is why so many otherwise forward-looking faculty resist it or don't manage to "get around" to it—because of implicit recognition that their epistemic and pedagogical power will be eroded.[1]

DOI: 10.4324/9781003037439-6

This erosion of power is necessary. If a pluriversal classroom is possible, it requires intensive and thorough work—to identify what it means to facilitate a classroom that fundamentally does not center western disciplinary knowledge.

In the past decades, and particularly in the 1990s, it was popular in pedagogical conversations to discuss "decentering" the instructor. In a decentered classroom, the teacher would not be the prioritized voice in the room. This has been a touchstone of some critical pedagogy scholarship, with Freire as a primary interlocutor. Critical pedagogy has made invaluable contributions, but it alone has not dismantled a colonialist system. Eve Tuck and K. Wayne Yang's analysis of how Freire's pedagogy reproduces colonial assumptions[2] is representative. The act of decentering the power-holder in the room is neither easy nor sufficient for dismantling the white supremacist reality of the university as it plays out in our classrooms.

At times, the conversation around "decentered" education has been reminiscent of what Tema Okun[3] identifies as part of a white supremacist framework: where "much energy [is] spent trying to prevent abuse and protect power as it exists." The promise of decentering the authority of the instructor has been suggested as mitigating the "gatekeeping" power of the instructor, without dismantling the systemic forces that reinforce existing hierarchies.

Other "remedies" offered in educational scholarship have functioned similarly: the proposition that transparency about grading processes equates to an antiracist practice, for example. While important, limiting the grading authority of the teacher will not alone make a classroom antiracist, and any other isolated pedagogical changes will not be sufficient in developing an epistemically just classroom.

In writing this, I in no way wish to claim that I have arrived at an antiracist, epistemically just practice in my classroom. My work has been primarily to say, "I notice. And I want to be accountable." This chapter is meant to call into question the conventions of university classrooms as they operate systemically. What follows is not a doctrine or program, but is meant to open questions we can collectively ask and respond to in our particular contexts. The goal is first to observe and then to come alongside our students in transforming the systems we are imbricated in. I don't want to fall into the shortcomings of positing a particular mode as a normative ideal; instead, I provisionally suggest that we develop the attention to recognize what is occurring moment-by-moment, coming to notice what is emergent and what is relying on the entrenched, harmful, and outmoded. We notice as our students do.

Institutional education is, in many ways, not good for us. With mental health crises, pervasive anxiety, and accumulated student debt,[4] the educational system as it stands is correlated with a range of deleterious effects. The system perpetuates ableism and classism in continuing to reinforce many forms of inaccessibility, rewarding forms of overwork that are possible for only those with the privilege to participate in this lifestyle. We experience educational harms first-hand, and we also witness their effects on our students. Nonetheless, we continue with education as it has been done. Meyerhoff describes this circumstance as an "addiction to education" as it stands.[5] Educational systems are held as a "good" despite the reality that racism, classism, ableism, cis-hetero-sexism, and misogyny are pervasive in them; these axes condition the health effects of education. To dissolve the culture that perpetuates these interlocking forms of oppression requires bringing into being a radically different way of showing up with and for each other.

Addressing white supremacy culture in education

What does this dissolution and emergence look like in practice? The collaborative list that was compiled by Tema Okun 20 years ago to name the characteristics of white supremacy culture provides a starting place for this work of bringing into being. Okun's list is organized into 15 major themes. I break these themes into smaller motifs in the following list, mapping them onto common educational conventions. I've named these motifs in terms other than those provided by Okun's document in order to further elucidate them. These motifs are overlapping and mutually reinforcing, and while Okun's focus is on racism, in particular, several of the attributes she describes also have been named in feminist, queer liberation, and disability justice scholarship.[6] These features are ways of naming the dominant epistemology of institutional culture (i.e., what is believed to be true about the human being within an institution and how that institution functions). The characteristics also indicate how epistemic injustice is perpetuated in academic settings. The following characteristics describe a range of institutions that have prioritized white people, reproducing exclusions and harms. When an institution is understood as "predominantly white," it is because whiteness dominates. Okun outlines the assumptions that this culture of domination mobilizes. The list that follows suggests how these assumptions translate to higher education:

- Shame: White supremacy culture includes the tacit conflation: "making a mistake is confused with being a mistake, doing wrong with being wrong." This orientation cultivates a sense that every day and moment is "high stakes" with the risk always that one can become deemed less-than-worthy. The deficit-based thinking that Okun describes orients toward humans' inadequacies rather than strengths. This reinforces a shame-based culture that is fundamentally disciplining. This culture of shame is readily internalized: with individuals accepting ideas of unworthiness and deficiency (e.g., "I'm not a good student").[7]
- Diagnosis: White supremacy culture emphasizes judgment and diagnosis. As Okun writes, "mistakes are seen as personal, i.e. they reflect badly on the person making them as opposed to being seen for what they are: mistakes." In the university setting, this may show up in how students are designated (e.g., "probationary") or the ways in which students are discussed and labeled (e.g., as a "C-student"), with little criticality about the assumptions behind these labels nor the stakes of labeling. Labels are fixed, negating a regard for error as both ephemeral and necessary for growth. This

orientation reinforces a hierarchical model of "knowing best" about another, without engaging them directly.

- Critique: Modes of critique assert a superior position over and against that which is the object of criticism. The culture of critique in Okun's words reflects a "tendency to identify what is wrong," with "little ability to identify, name, and appreciate what is right." In classroom spaces, meeting spaces, and throughout academe, relatively little time and energy is dedicated for appreciation and celebration; instead, the focus is on lack. In this, the inadequacy of human beings is the focal point, rather than the inadequacy of the systems that shape their behaviors.
- Fixed Mindset: Despite talk about fostering growth mindsets in our students, the conventions and structures of academia often do little to put a growth orientation into practice, as summative evaluations are prioritized over supportive and formative processes. Okun observes of white-dominated institutions: "little time, energy, or money put into reflection or identifying lessons learned that can improve practice, in other words little or no learning from mistakes." Feedback on classroom practice is sought once or twice per term, and typically only at the end of the term. There is little opportunity for the co-construction of an educational space that is reflective of, and responsive to, the community that composes it.
- Conformity and binary thinking: The fixed mindset described previously is an outgrowth of binary, "either/or" thinking, which Okun defines in this way: "things are either/or, good/bad, right/wrong, with us/against us." An ability to hold multiple "truths" at once is not cultivated (i.e., "both/and" orientations). Complexity is undervalued, as is the presence of many alternatives. Obedience to established conventions is insisted upon; when agreement is not present, someone is to blame and made to be "other." Inclusion is translated as "you're welcome here as long as you are like us."[8]
- Defensiveness: People respond to new or challenging ideas with defensiveness, making it very difficult to raise these ideas and more so to create systemic change. There is a general sense of resistance to change. Defensiveness is catered to, to the extent that, in Okun's words, "a lot of energy in the organization is spent trying to make sure that people's feelings aren't getting hurt or working around defensive people." This might include students and colleagues who demonstrate white fragility or who deny, dismiss, or denigrate testimonies of harms—both in terms of the learning content (e.g., which histories and accounts are learned in the content of the course material) and in terms of what transpires between people in educational spaces (e.g., how the racism that is evident in the here and now is or is not addressed).
- Conflict suppression: Conflict, or the tensions that necessarily come with difference, is neither expected nor welcomed, and thus difference is not fully accepted. When department or classroom cultures pride themselves on being "conflict free," difference is being covered over in the calcification of established norms. In this circumstance, conflicts are suppressed or displaced, often

through the blaming and shaming orientations described earlier. Conflict is prevented from being generative.

- Paternalism: Institutional culture rejects the ethic expressed in the slogan: "nothing about us without us is for us." As Okun defines it, paternalism means that "those with power think they are capable of making decisions for and in the interests of those without power." This is a basic premise of education, as syllabi are designed *for* a group of people that an instructor has never met, let alone consulted. For those with decision-making power, transparency is not required. Consent is not sought.
- Singular mind, singular power-holder: As Okun notes, there is little, if any, regard for the value of sharing power. This is evidenced in institutional barriers to co-taught courses, shared leadership positions (co-chairships, co-deanships, co-presidencies), etc. The "singular genius" orientation is remobilized in institutional conventions, and Okun notes, "those with power feel threatened when anyone suggests changes in how things should be done in the organization." Suggestions for change are regarded as "a reflection on their leadership." This individualist orientation is related to a culture of competition.
- Disciplining through constructs of "civility": Testimony about harms or the questioning of norms is silenced through the ascription of what is "rude" or "inappropriate." These terms—and with terms like "civility" and "tolerance"—protect the status quo. Such terms are utilized but not defined. Those who count as belonging to the inside of a system are entitled to being read as "civil," while others suffer exclusion through the weaponization of "civility" and related terms.
- Productivity models: It is assumed that there is value in being "data-driven." At an institutional level, refining the measurability of learning outcomes takes precedence over students' experiences of the institution. Faculty quantify their work through tenure files; they quantify students' work through point systems and grading schemes. The qualities of human experience and process are overlooked. Measurable goals are most highly prioritized.
- Planning over presence: A designated "plan" is prioritized (e.g., moving through the agenda or the lesson plan) over and above the need for responsiveness to what is present in members' experiences. Okun observes that white supremacy culture includes "no understanding that when there is a conflict between content (the agenda of the meeting) and process (people's need to be heard or engaged)" people's needs must ultimately be prioritized. Okun writes, "if you haven't paid attention to people's need to be heard, the decisions made at the meeting are undermined and/or disregarded." Even still, allegiance to the (often written, often single-authored) plan is the default.
- Expertise over experience: Working in parallel with what I am calling the "planning over presence" orientation, the prioritization of expertise over experience privileges content over human-to-human relationality. In institutional culture, if an experience doesn't come wrapped in a credentialed theory, with citation of accepted forms of expertise, it is invalidated. Testimony goes

unheard unless it comes packaged in sanctioned reference points. In turn, a closed circle of "expertise" is maintained.

- Supposed objectivity: As Okun writes, "the belief that there is such a thing as being objective" is a pernicious myth in institutional culture. Against this belief, emotions are understood as "inherently destructive, irrational, and should not play a role in decision-making or group process." This is a mechanism of invalidation. Those whose emotionality becomes visible (because their emotional reality does not reflect that of those who are in power) are rendered uncredible. This consequential bias is related to the zero-point orientation or the assumption that knowledge can come from a neutral space outside.[9] Within a zero-point orientation, Eli Meyerhoff explains, "expert knowledge is seen as universally valid, from a position above any particular bodies and places in the world." Objectivity denies the situatedness and embodied nature of knowledge and denigrates those who do not fit within the privileged space of the "zero point."
- Prioritizing the emotionality of the privileged: The emotional experience (i.e., the comfort) of power-holders is prioritized, even as they may deny the presence of emotion and instead use emotion as a means of othering. Those in positions of privilege demand what Okun calls a "right to comfort," which translates to blame of those who do not share in the status of comfort. The blame is rendered sometimes with terms that displace the emotional realities of the privileged onto those they silence: the "feminist killjoy" is blamed for killing the joy; the person with anger is invalidated and made to be a mere nuisance to the otherwise comfortable waters of those who benefit from the status quo.
- Sense of urgency: Moving rapidly keeps the status quo in place. Doing what is most efficient often means doing it as it has been done in the past, thereby repeating historically marginalizing operations. This culture does not facilitate the slow, deliberate work of transformation. Instead, time becomes commoditized and used to reproduce what is already institutionalized.
- Myth of meritocracy: Established hierarchies are unquestioned and assumed to be based on a fair and equitable meritocratic system. This is related to what Meyerhoff calls the "vertical imaginary" of higher education: students are expected to move *up* the levels of schooling and then through the ranks of a profession. In Meyerhoff's terms, this vertical imaginary is intertwined with a romantic narrative that "students face obstacles, and overcome them as heroic individuals." This obscures the tremendous violence of these obstacles and how they often cannot be decidedly "overcome." The insistence on meritocratic justifications for existing hierarchies creates obedience to them, while suppressing that which would prompt resistance toward them.

Opportunities: shifting white supremacy culture

Based on many forms of epistemic injustice described earlier, what does it mean to counter white supremacy culture? It is not enough to critique what is; what values can point us toward a more epistemically just academic environment? Following from the "antidotes" suggested by Okun, as a way of countering the force of white supremacist assumptions, the following list offers a few starting points. The values expressed in what follows can be read against the previous discussion, demonstrating what it means to shift from supremacist, hierarchical practices that uphold white supremacist capitalist colonial ableist heteropatriarchy.

- Appreciation: In the discussion of perfectionism, Okun notes that a characteristic of white supremacy culture is a lack of expressed appreciation for each other. How often do we express appreciation for students' work? How often do we encourage students to express appreciation for each other (during small group work, class sessions, or other class rituals)? How is appreciation a central, meaningful—and not perfunctory or performative—practice in all the work we do?
- Growth and learning: As Okun notes, what characterizes an anti-oppressive learning organization is the rejection of shame-based perfectionism that prevents flexibility. If growth and learning are prioritized, humility and suppleness follow; these in turn allow for change. Plans and policy statements should be understood as "living documents," available for adjustment as collective learning continues. These structural considerations reflect members' personal orientations toward continued responsiveness.
- Receptiveness: Those in power need to be deliberate in approaching issues with receptivity, establishing pathways for feedback to be shared in ways that do not add taxation or pressure for those who offer their experience. A promise to listen without defensiveness is a starting place for accountability. As Okun writes, "don't require those who raise hard issues to raise them in acceptable ways, especially if you are using the ways in which issues are raised as an excuse not to address the issues being raised."
- Embrace of change: Change should be actively sought, with full acknowledgment that the systems as they have been structured have perpetuated harms and have actively excluded marginalized groups from the levers of power. An appeal to how things have been traditionally done is an appeal to keep these systems in place.
- Diversity in many forms: To seek diversity with a justice framework means not to tokenistically recognize, but instead to push aside those barriers that are preventing diversity from being respected or included. Diversity includes racial, gender, and class-based diversity along with linguistic diversity, neurological diversity, affective diversity, epistemic diversity, and innumerable other forms. As Okun notes, "push yourself to sit with discomfort when people are expressing themselves in ways which are not familiar to you."
- Shared decision-making: Resisting the paternalistic orientation of the institution requires that those who are affected by a decision are brought together from the start to shape and co-construct a vision. Proposals require consent of all involved, interested, and affected parties. This process takes time, countering a culture of urgency. The art of collaboratively seeking mutual understanding

and power-sharing should be explicitly discussed and continually improved. Respect means fully acknowledging what each stakeholder brings to the table in terms of knowledge, skills, values, and concerns. Shared decision-making also entails information-sharing, and this cannot be limited to written documentation. Honoring many different modalities of expression and record-keeping is necessary.
- Consent: Freedom and choice should be actively built into organizational structures as a way of countering the disciplining force of institutional practice.

This initial statement of principles and characteristics can provide the foundation for rethinking teaching practice.

Contemplative practices can align with and support many of the "antidotes" listed here, as these practices allow us to notice what is present, to accept without attaching, and thereby make space for change. Some mindfulness practices may emphasize gratitude and appreciation, as they also emphasize connection—a shared sense of interdependence. In this, contemplative practices can make space for the imperfect. Contemplative practice meets the presence of pain, rather than suppressing it—being with pain in order to heal it. At the same time, contemplative practices can be steadying, in contrast to the pursuit of control. The steadiness of contemplative practice can translate to willingly and courageously embracing change.

Contemplative practice can provide the scaffolding to support the internal reorientations that are required to counter the oppressive culture that Okun describes. While of course not every facilitator of a classroom space will have a robust mindfulness practice, even the simple act of slowing down enough to observe and question may be key to creating change. Slowing down, noticing, and finding internal resources for change are all necessary. By observing with intention, norms, once invisible, can be defamiliarized and called into question.

The class without a syllabus

We do well to pause on each convention of the classroom, to interrogate it for what it represents. Consider the syllabus, for example. As a genre, the syllabus is a presentation of the terms and topics that are set by the instructor for the course; the syllabus is *assigned*. This genre represents a theory of education that says that leaning can be, at least in part, pre-planned and programmed. The linearity of the syllabus structure indicates an idea of progression that, it is assumed, all who come to a course will fall into step with. Any diversity of thought, if allowed, will ultimately be folded into what is outlined in the schedule. Growth and learning are legible only insofar as they are anticipated in pre-written learning outcomes.

A syllabus may be generously read as a set of agreements between instructor and student, stated explicitly at the outset of the term and utilized for future accountability. But often this accountability is unidirectional, with the teacher having far more freedom and decision-making power than the student. The syllabus is a symbol of the one-mind approach to pedagogy, with the teacher as author. Any shared process in the classroom will be subsumed under the pre-determined course structure and the accompanying pre-built "learning management" website.

The problems with this model are many. It is profoundly reductive to prioritize one mind over all others in the room, even if that mind has had years of training. That training offers something to the group of people who come together under a course description; that training is necessary and of course should be invited in, but it represents only part of what can be shared in the space. A classroom cannot fully activate the cultural wealth[10] of a student if that student has not participated in the basic decision-making that creates the educational situation.

What would it mean to write a letter of invitation, rather than a syllabus—a letter that may still propose a learning journey, but in ways that invite co-creation? A proposal is something far different than a determined outline. A proposal may allow for an educational experience that is a collaboration, a partnership, a negotiation.[11] A proposal can offer direction in a way that is deliberately responsive and relational; it can enact an idea of learning that is focused on *connecting with* ideas and people, rather than *distributing* pre-given content.

To prioritize the cultural wealth of each student who comes to a class does not mean that the instructor in the room has no role to play. The disciplinary knowledge can be brought in and attached to what students offer to each educational space. Starting with the students' interests, the exigencies that matter to them, and the cultural beliefs and investments that they want to preserve creates a ground for the educational endeavor. Knowledges can be added from here. Reading lists can be suggested, accompanying the questions: *What do you want time to read? What reading activity do your projects ask for? What interlocutors will support and broaden your thinking?*

With the role of expanding students' awareness of what is possible and what has been said, the teacher is positioned as a facilitator[12] who is equipped to say: "have you considered this?" when that question does not carry the weight of a "must" (as in "consider this in order to earn an 'A' from me"). The instructor's knowledge is infused into the students' educational journey in a format that honors the students' experiences, cultural orientations, and choice.

I think of the model of the literature table, utilized by the Combahee River Collective, that prioritized resource-sharing.[13] Members who came to the space for dialogue would place facsimiles of readings onto a table to be browsed and shared by anyone who had interest. What if our classrooms followed this model, with students resource-sharing with each other alongside their instructors? We can pose the question: *What are the texts that you feel we must read? What texts can we read with you, to help us understand your project? What texts help us to think about, and act on, that which we collectively care about?*

We can consider the breadth and range of cultural orientations toward central issues by first prioritizing students' situatedness and the knowledges and interests they bring to the classroom space. In seeking a culturally affirming practice that honors marginalized knowledges, this pedagogy pushes back against the echo-chamber of our algorithmically curated and mediated worlds. The students' knowledges and interests grow through the encounter with voices and ways of perceiving the world they have not experienced before. The instructor's role is to help students keep pushing beyond what they already know, learning their own epistemic humility, but learning this from a place of groundedness in their cultural orientations and interests. Humility emerges out of affirming spaces; finding security in one's own identity may be necessary to finding humility. The instructor can also help students to hold the "both/and" of what they encounter, the complexities and layers and irreconcilable realities they find.

This method prioritizes choice and consent and relationality between students and their teachers. With such an orientation, this method pushes against the tendencies of an educational system that is more and more reliant on predetermined, readymade forms of education: courses that have been pre-built in content management systems with the same interfaces as the massive open online courses (MOOCs) that impersonally see thousands of students churn through pre-given content. In the MOOC theory of education, responsiveness to each individual student is not only impossible (with such "massive" numbers) but is not part of the epistemology that undergirds the educational design. The teacher creates the modules, and the students move through them. The user experience (UX) might be considered, but responsiveness is flattened into content or interface tweaks, not human-to-human vulnerability.

In this context, relationality may be the most salient and important offering of the brick-and-mortar and live virtual classes. In an educational marketplace where content is readily available, teaching with radical relationality may be our most important way of responding to what education is becoming.

A relational pedagogy trusts in a person's desire to learn. There is pleasure in learning when it's the learning that is sought, and we can't know in advance what learning each student seeks. Of course, that desire to learn can be easily crowded out by life demands, caretaking responsibilities, material concerns, and the ensnarements of the economy. The educational space needs to be a protected space, a space where students can be supported in pursuing that which transcends conditioned decisions of where to place their attention. Meaningful work is challenging and there are many reasons why one might choose to neglect it day-to-day, but the classroom can be a space where students are called back to their original intentions toward meaning-making and meaning-finding through environmental and humanistic study.

What if we measured our success as educators not by an input/output model of skills and concepts acquired, but by students' felt sense of inclusion, wellbeing, inspiration to learn more, confidence to address central problems, and the feeling of cultural affirmation in a classroom setting? I pose this question based on Michi Saagiig Nishnaabeg scholar Leanne Betasamosake Simpson's vision and call: "We need to join together in rebellion of love, persistence, commitment, and profound caring and create constellations of co-resistance, working together toward a radical alternative present based on deep reciprocity and the gorgeous generative refusal of colonial recognition."[14] What if coursework meant doing the work Simpson describes? Simpson envisions practices that provide a "flight path out of settler colonialism" and that don't "endlessly accommodate whiteness," don't "accept the inevitability of capitalism," and that mark a "disruption to the hierarchy of heteropatriarchy."[15]

Noninterference, self-determination, autonomy, and freedom

Simpson's book *As We Have Always Done* underscores the necessity of self-determination and sovereignty in education. She writes,

> I am absolutely sure that we as Nishnaabeg cannot survive as a people without creating generations of artists, thinkers, makers, and doers that live in Nishnaabeg worlds, that are in respectful relationship with each other, that create a movement that joins us to other Indigenous nations to protect the land and bodies. We need to live deliberately and with meaning.[16]

Acknowledging Neil Roberts' definition of marronage—"a group of persons isolating themselves from a surrounding society in order to create a fully autonomous community"[17]—she uses the term, derived from Arawak and Tainos thought, to note the importance of self-determination. She writes, "*how* we live, *how* we organize, *how* we engage in the world—the process—not only frames the outcome, it is the transformation. The *how* molds and then gives birth to the present. The *how* changes us. *How* is the theoretical intervention"[18] into colonialism, and its continual attacks. She continues, defining colonialism:

> I understand colonialism as an overwhelmingly dominating force in my daily life that continually attacks my freedom and well-being as kwe. Colonialism tries very hard to keep me off my land. It tries very hard to ensure I cannot speak my language, think as my Ancestors did, find comfort in elders or the river or the lake of rice. It tries very hard to get me to think in a particular way. It tries very hard to get me to resist in a particular way. It tries very hard to get me to move my territory in a particular way. It controls how I make a living and how I feed my family. It tries to control the relationship I have with my children. It tries to control my sexuality, the ways I express my gender, how I take care of myself, and how I parent with escalating magnitudes of structural and interpersonal discipline and violence if I do not conform. It creates a world where I am never safe. It is a violent system of continual harm forced on my body, mind, emotions, and spirit designed to destroy my ability to attach to my land, to function as kwe, to be a grounded, influencing agent in the world. I am interested in freedom, not survival, and as kwe, I understand my freedom is dependent upon the destruction of settler colonialism.[19]

How one lives, how one approaches education, how one dwells with another—all can be a counter to the destruction and entrapments of colonialism. These are the central concerns of an educational practice.

Within an anti-colonial educational system determined by Nishnaabeg peoples,

> there is no standard curriculum because it is impossible to generate a curriculum for 'that which is given to us lovingly from the spirits,' and because it doesn't make sense for everyone to master the same body of factual information. Nishnaabeg society in its fullest realization requires a diversity of excellence to continue to produce an abundance of supportive relationships.[20]

This is what is meant by diversity and relationality in education. There is no one teacher-decider. Indeed, "it is unthinkable within an Nishanaabeg worldview for a leader to impose their will on their people." In parallel, "it is unthinkable to impose an agenda onto another living thing—in essence the *context is the curriculum*, and land, Aki, is the context."[21]

The singular great mind who "knows best for" another human, plant, or species is a colonial formation inherited in our educational system. It eradicates reciprocity and attentiveness to another, to land, and to understanding of self in relation to multispecies community. This is the dispossession of colonialism. "Within Nishnaabeg thought," Simpson writes, "the opposite of dispossession is not possession, it is deep, reciprocal, consensual *attachment*."[22] That colonialism is a structure and not an event[23] means that this dispossession, this violence against Indigenous peoples' onto-epistemologies and relations, is continually experienced, and it manifests in continued colonialist education.

This is the dehumanization of colonialism as it strips the dignity of relations. As Linda Tuhiwai Smith notes,

> The whole process of colonization can be viewed as a stripping away of *mana* (our standing in the eyes of others and therefore in our own eyes), and an undermining of *rangatiratanga* (our ability and right to determine our destinies).[24]

This is what is at stake when we speak of epistemic justice: the very reality of what it means to live, to have one's existence be recognized, held as sovereign.

Epistemic justice requires the preservation of self-determination—which entails resistance to the global white-supremacist regimes that would encroach upon it. Macarena Gómez-Barris calls for an understanding that does not stop at resistance, but sees the "importance of epistemological autonomy and embodied knowledge as necessary to pushing away from a paradigm of mere resistance into the more layered terrain of potential."[25] To work on environmental issues is to work within this context of struggle.

One thinks here of the autonomous zones established by Zapatista community members who

> affirmed their commitment to do things themselves based on their own ideas and with their own resources. . . . There was the assumption among many sympathizers of the EZLN that autonomy was by its very essence radical in its unsettling of state projects.[26]

In her research living and practicing solidarity in Diecisiete de Noviembre, a rebel municipality named after the date the EZLN was founded, Mariana Mora writes of hearing the comment made again and again: "Autonomy is for life; it is about the possibility of a new life." The Tseltal word for "life" is kuxlejal, "life-existence." The term is embedded in concepts like *stalel jkuxlejaltik*, a way of being in the world as a people, and *lekil kuxlejal*, the daily aspiration to live a good life in relation to the earth, the natural and supernatural world, and all living beings. Interrelatedness is prioritized through autonomy, the right to preserve relational lifeways and knowledges.

The education that we seek to practice in the humanities should be deliberately directed toward the honoring of sovereignty. This means not only eradicating the colonialist, whitewashed histories that are still readily circulated; it also means working in solidarity with movements that counter assimilationist forces in a neoliberal world system. To hold space for pluriversality and the right to self-determination is not to welcome an "anything goes" curriculum. To believe in the necessity of pluriversality is to have a stance in the classroom, and this stance can only be rendered in integrity if the mode and form of education allow for plurality.

What is needed is a pedagogy of indispensability that, in David Naguib Pellow's definition of the term, "honors key environmental justice and ecological principles by seeing all communities (more-than-human and human) as interconnected, interdependent, but also sovereign and requiring the solidarity of others."[27] We can interrogate dominant ideologies, while also providing a sense of interdependence in a radically plural world. In the anxiety of a climate-changed reality, we seek imaginations that work against oppressive regimes and build responsive multispecies futures. Developing this cultural imagination requires standing in the "not knowing" of ecological and pluriversal complexity—recognizing that all ecologies and experiences of being are beyond quantification. No ecology, and no one being, is ever fully transparent and knowable to any of us; if we make a practice of remembering this in humility, how might our day-to-day pedagogies change? How can we seek an education that restores and cultivates interiority and relationality?

What would it mean to find a way of practicing the environmental humanities that honors the knowing practices of bodyminds across human and more-than-human interactions? Mindfulness practices reveal to us the ways we have of knowing the world—ways that are not predetermined or programmatic. Through attention, we come to know our fellow transcorporeal[28] beings of this planet, across species, with greater respect.

Classroom as community

There is power in acknowledgment of an interdependent community, and there is power in gathering. There is power in coming together in response, but we know that this work of coming together is complex. Foregrounding community requires careful and deliberate facilitation.

If the classroom can be a lab for solidarity building, we must remember that solidarity is hard-won. As Dina Gilio-Whitaker notes, between us there are "critical ruptures that shouldn't be overlooked so that we can 'all just get along.'" There are "cultural differences and unspoken assumptions that often still interfere with non-Native activists' ability to be respectful, effective allies and accomplices."[29] There are long, fraught legacies that position us in tension with each other. What does it mean to build together, to counter the relentless us/them divisions of authoritarian regimes?

The classroom is a space of coming together, but often it is not leveraged in that way. With individual grades handed out to students who do individual projects, common classroom practices rest upon an idea of the isolated and atomized self. In so doing, they miss the power in community and the potential to work through some of the most complex questions of what it means to dwell alongside one another, to be together in struggle, to honor and respect differing regards for, and experiences of, ecological issues.

I often start my classes with one of the principles of emergent strategy from adrienne maree brown. I share with my students brown's words: *There is a conversation in the room that only these people at this moment can have. Find it.*[30] Based on this principle, brown asks a group to arrive at answers to the following questions: "Why are we meeting? What can this group uniquely accomplish?" She reflects on these questions,

> There are always a ton of relevant conversations that could happen, but there is usually a very small set of conversations that a particular group, at a particular moment in history, can have and move forward, given their capacity resources, time, focus, and beliefs.[31]

The idea is to bring students together under a sense that the most interesting and important parts of their educational journey can come from what they do in community. This orientation values what each bring to our discussion table and to the world beyond the classroom—this to counter the expectation that they're here to focus on me as their instructor. In the classroom, I, as a facilitator, am a part of a web of relationality, but I am not central to it. I invite students to enjoy each other's company. As facilitator, my role is to help a group come to trust each other and to honor the trust that is being given to them. *Trust the People. If you trust the people, they become trustworthy*, is another principle of emergent strategy. adrienne maree brown encourages us: *Move at the speed of trust. Focus on critical connections more than critical mass—build the resilience by building the relationships.*[32]

Deliberateness is required in building a community. Priya Parker is a professional facilitator who has theorized the art of building community in her book *The Art of Gathering*. She describes her work, "My job is to put the right people in a room and help them to collectively think, dream, argue, heal, envision, trust, and connect for a larger purpose."[33] Parker emphasizes the importance of gathering with purpose, and by *purpose* she means something far different from a learning outcome. To have a purpose for coming together is not to have a predetermined idea of what ideas participants will walk away with, what concepts they will have accepted and come to believe by the end; instead, a purpose is a statement of what might become of a gathering, why the gathering has a unique potential, and why that potential matters. Similar to the teaching principle: "never ask a question that you already know the answer to," the idea of a purpose, in my reading of Parker, is to set in motion that which you don't already know the outcome of—to pose a hope that cannot be guaranteed, that will be shaped by the people who come to it.

A strong purpose gives momentum. In the environmental humanities, we might gather in order to address a local issue, to strategize interventions, to actively and deliberately build solidarity, to find forms of support in each other for the anxieties and traumas we are differentially facing in a climate-changed world, and so on. We might gather to reinvigorate a sense of wonder and enchantment with the natural world. We might gather to push out of a conceptual logjam or to enliven a radical imagination for radical change. Each of these purposes leads to a different form of being together, of *making with*. The gathering looks and feels different under each of these purposes. If we are gathering to respond to a corporation's proposal to place a wind farm site on Indigenous sacred ground, our time might be spent identifying rhetorical and activist strategies, considering the local political landscape, researching legal precedence, reading about sacredness as an environmental justice issue, and so on.

As Parker notes, the purpose of the meeting is not to attend the meeting, the purpose of the class is not to attend class. The purpose of class time is not to relay information; this is not an effective use of the power and potential of gathering. The purpose is to come together for a reason and to contribute together. Parker encourages a *vivid* and specific purpose for each moment of gathering: What is the purpose of this class session on this particular day, for the people who have come to the space at this present moment? What does it mean to make this class session a source of growth, support, guidance, and inspiration for this particular time and place? What are we here to do together, and why do we need each other for this? Why does this matter to us? What needs does this coming together serve? What problem might be solved by coming together? What do we want to be different because we have gathered? We are seeking a specific purpose that all who come to the classroom space can see themselves in and can care about.

Coming together in the classroom is a type of ritual. At the start, when a new group comes together, practices are established that will be repeated. How a class session begins, how it ends, how time is allocated, how contributions are acknowledged, all of these decisions shape students' experiences of the space. The establishment of these practices should be done thoughtfully and in alignment with the purpose of the gathering and the intention to create a space of vibrant exchange, conscious of power dynamics in the space. Each choice communicates messages about the values, power relations, and assumptions of the space, so each choice should be analyzed. For example, compare the choice to begin with a meditation or moment for reflection led by a different member of the group each day with the choice to have the teacher start a session with a reminder of what that instructor has assigned and the accompanying due dates that are required.

The repetitions come to form a group's sense of collective identity. As Parker notes, rituals say: *This is who we are. This is the way we do things around here.*[34] Because the repetitions of the classroom have this effect, they have the potential to communicate belonging as much as they can produce exclusion. Co-creating the conventions and agreements that will shape the time with all members of the classroom can be one way of mitigating the risks of exclusion. For example, the question might be posed to students in the room: What do you need from each other in order to feel comfortable bringing your full self to this space? This question might be posed after an initial exercise (such as a short free-write response to a prompt), asking the group: what would you need to feel comfortable sharing your writing with the people in this room? What can we agree to, in order to make everyone in this room feel ready to be a little bit vulnerable—knowing that work that we care about is work we are willing to bring our multidimensional selves too?

adrienne maree brown's texts can be an important reference point for these discussions regarding what we need from each other and the agreements we would like to share. Her books *Emergent Strategy* and *We Will Not Cancel Us* open up conversations about what it means to come together, what it means to establish a fruitful practice that reflects an antiracist, liberatory vision. These books account for a range of considerations: trust-building, addressing harm, shared decision-making, agenda-setting, tracking progress in ways that subvert linearity, and so on. brown suggests ways of addressing conflicts that do not fall into the politics of disposability. brown's books also address the importance of listening with presence and mindfulness. brown writes, "To a certain degree, our entire future may depend on learning to listen, listen without assumptions or defenses."[35] Listening means, in brown's clever framing, "relinquishing Frankenstein": She writes,

> You are not creating people to be with, or work with, some idealized individuals made of perfect parts of personality that you discovered on your life journey. You are meeting individuals with their own full lives behind and ahead of them. Stop trying to make and fix others, and instead be curious about what they have made of themselves.[36]

This is the posture we should take in relation to our students, and we can also hope to participate in creating the conditions where students can find this relationality with each other. This prospect requires overcoming the anxieties and divisions that structure our society. We can give language to these realities, offer strategies for overcoming the ways in which social distrust may be structured into our relations, and invite students to co-create the world they would like to see by being together in the classroom.

The act of building community is difficult work, but it may be one of the most essential parts of what an educational experience can bring. This learning can be scaffolded through texts like the "Principles of Working Together," adopted at the Second People of Color Environmental Leadership Summit, held in Washington, DC on October 26, 2002. Statements of coalition like this are important texts for study, as we seek to make the classroom a coalitional space.

Transforming what education means in these ways is a disruption of the status quo. As with any disruption, we can anticipate forms of discomfort and entrenchment. We can give students explicit language for naming their experience as they encounter voices that splinter dominant epistemologies. We can help students notice their investments in master narratives, with cognizance that, in Kevin Kumashiro's words:

> the problem that educators need to address is not merely a lack of knowledge, but a resistance to knowledge, and in particular a resistance to any knowledge that disrupts what the students already know . . . unconsciously, students often want learning that affirms their knowledge and self-identifications. In particular, students often desire learning that affirms their belief that they are good people and resist learning anything that reveals their complicity with oppression.[37]

When complicity with oppression is noted, the fear of being "canceled" may show up. We can help students notice how cancellation too derives from colonial logics that link validity with purity. It can be important to name this and other fears that are present and to find ways of agreeing to practices of accountability that enable everyone to know where they stand in relation to the community.

Participants in a classroom community need to know that they can trust the facilitator to offer pathways for these difficult discussions, that the facilitator will not shirk the need for candor, and that the facilitator can be clear-eyed and ready to address power dynamics in the room. Being clear and transparent about what I, as facilitator, am responsible for is a key aspect of this trust-building process.

The role of the facilitator

The facilitator is responsible for creating a space for shared work; providing multiple pathways to gain the input of all group members; and offering proposals for how the group might move forward next based on shared feedback.[38] The facilitator reminds the group of agreed-upon purposes and principles. The facilitator also infuses resources (e.g., readings, information, and tools) that speak to what is emerging in our conversations, inviting multiple perspectives into the space. While the participants are most central to the process, the facilitator's role is consequential.

The facilitator is in a sustained and active relationship with each participant in the room, learning from them and responding to what emerges. The facilitator is also active in creating a culture where everyone can learn in ways aligned with a shared purpose of moving toward a more antiracist, anti-oppressive, and transformative world.

As another part of the role, the facilitator is attentive to pacing, time constraints, environmental issues (i.e., accessibility in the set-up of the room, ensuring that everyone has a balanced position in a shared space, etc.), and the necessity of carving out time and space to reflect and offer meta-level feedback on the process. The facilitator might ask at the beginning of a class session: "What do you need from this meeting?" Correspondingly, at the end of the class session, the facilitator might ask: "What are you noticing? Based on what you observe, what do you think this group needs next?" adrienne maree brown suggests a question from Gibrán Rivera: "What is the next most elegant step?" She explains,

> An elegant step is one that acknowledges what is known and unknown, and what the capacity of this group actually is. An elegant step allows humility, allows people to say, "Actually we need to do some research" or "Actually we need to talk to some folks not in this room" or "Actually we need a full day to build this plan out into something realistic and attainable."[39]

As facilitators, each choice we make is shaping and requires careful consideration. For example, think of the language we use: Is the language invitational and humanizing, or does the language command and reinforce power structures? Even the term "assign" can be read in these terms: do we *assign* work, or do we invite students into a process that they are ultimately in charge of? Do our choices work to create a protected space (even as it cannot be a fully safe space) for students to do the learning they want to learn, in relation to each other?

Transparency might be what is most needed from us as facilitators. We can offer candor about the rationale behind our choices. Transparency can be part of accountability; it also entails humility about the ways in which our choices may fall short of our ideals. As Berila notes, "The more honest I can be about the progress and missteps I have made, the more it seems to give them [students] permission to be with their own journey, whatever that may be."[40]

Vulnerability from the facilitator can prompt vulnerability from the participants.[41] adrienne maree brown suggests that the active practice of vulnerability can make the connections more "clear, open, available, durable."[42] This entails destabilizing the divide between teacher and student. We become peers of each other.

Mindfulness opens a way of grounding our considerations in terms of our humanity, our embodied relatedness, and our interconnectedness. It opens space for what we can observe of our full selves in a community context. Mindfulness also coincides with a pace that is slowed enough for inherited conventions to be questioned and rethought. This slowed pace allows for greater attentiveness, to notice what assumptions and scripts appear in the mind, to find the temporal distance from those assumptions enough to call them into question and find cognizance of all one does not know.

Such negotiations take time and patience, and it is important to prepare students that a different model of education is being introduced and practiced. Because students are accustomed to the (more "efficient") conventions practiced elsewhere in the institution, they may find themselves frustrated with a different model, especially those students who have benefited from the institutional practices as they are. As facilitator, the instructor can create avenues for finding language for experiences of frustration, anxiety, or concern about what this change in practice might mean. A discussion of how the formal and structural choices of the class reflect the values of relationality, epistemic plurality, and so on is necessary to foreground.

Some of the frustration, anxiety, or concern may stem from the reality that the teacher is still imbued with relative institutional power and that assessment practices are not suspended in the space. Ensuring instructor accountability regarding grading and assessment is a necessity. This may mean starting with a discussion prompted by questions such as:

> What did you come to this class to learn? How will you know you've learned what you need to learn? How can you make space for what you don't yet know you need to learn? Given that we are embedded in a system where grades matter, what approach to grading would serve you?

Suggestions of labor-based models of grading may be appropriate here.[43]

A balance can be sought between what is shared in community (e.g., collaborative projects, shared readings, etc.) and what is carried out by each student individually. Decisions about the work that the learning experience entails are rightly informed by each student's background and vision for the future. For some classes, students could write their own list of assignments they would like to complete, get feedback on, and revise, with the grade being earned from the completion of these steps. This would create space for students to experience the art of designing projects that are meaningful to them. *What is a "dream project" you would like to complete some day?* We might ask our students. The time in the classroom is precious time to move toward those intentions.

As Eli Meyerhoff has noted, "different modes of study are bound up with different modes of world-making."[44] We can practice ways of building a world in which many worlds fit within an educational space, as part of a long concatenation of visionary, coalitional work.

An artifact from the classroom

I close this chapter with the note that I offer students on the first day of class. I share this artifact not as a model, but as a form of reflection. The letter published here is one I have iterated on over the course of several years. I will continue to iterate on this letter with each class I teach, knowing that this initial means of welcoming a group is formative of what the community builds together in the weeks that follow.

Welcome!

This course is yours, collectively; it will be built by you and the colleagues you are meeting today. These are people who will be by your side as you continue on your path. I hope that your relationship to each other and your relationship to what we will be discussing in the coming weeks will continue long after our semester comes to an end.

As we move into the semester, I want to share some things I hope for. When I say "I hope for" these things, I mean: I am committed to doing what I can to make these things happen for us, but all of these things need your active contribution too.

I take my role as facilitator seriously and humbly. I will be learning alongside you and doing my best to respond to what I hear from our group, making suggestions based on what I understand of the group's needs. I hope that you'll keep an open dialogue with me to help me better understand how I can best serve our group in this role. That said, the work of building and sustaining a group—the kind of community space that we love to come to—is work that, by necessity, involves everyone.

- ***I hope we'll be in community together this semester.*** *You are a member of a community in this course. I hope you'll look for ways to actively contribute to the community in ways that feel meaningful to you. Let's do what we can to support all of the people who share our course.*
- ***I hope we'll consciously demonstrate respect for our differences.*** *Recognize that we come to this space with different backgrounds, assumptions, and orientations. This course will give you the opportunity to see yourself in a larger context and to grapple with your own background and assumptions. Note where there are gaps in your understanding. Be open to humbly asking questions. Our views are reflections of the worlds we have lived in; these views can change. Don't freeze up. Stay with the breath. Find the support of the earth beneath you as we move through this next part. We are here to learn. We are each on our own journeys. If we're doing things well, we will learn things about ourselves in different ways.*
- ***I hope you'll trust me enough to let me know what I have missed.*** *We are doing this self-reflexive work together, so please let me know when one of my awareness gaps appears in my facilitation. You can trust me to be receptive to what you have to say and to do my best to be accountable and to try to repair. If there are things I can adjust to make it more possible to put your full self into your education, I hope you'll let me know.*
- ***I hope we'll build trust with each other.*** *We all are on a learning journey. There are things we each know and things we each don't know. I hope we can trust each other enough to be able to learn about the unintended effects of words or actions. I hope we can trust each other enough to be able to share about those effects—how words and actions are received. There are false and harmful ideas that are "in the air" around us, and we may not realize that we have breathed them in until they show up in a space like ours. One of the things we can do for each other is to learn to be more self-reflexive and*

attentive to what we do, say, and hear—considering how micro-actions are related to macro-change.

- ***I hope you'll help me better understand how I can best support you.*** *You are welcome to make suggestions and requests. Please offer an alternative that would work better for you. I will be open and receptive to understanding your needs as an individual in the course and to strategizing with you how we can work together to meet those needs. I am committed to improving accessibility; if you discover something in the course that is not as accessible as it could be, please let me know as soon as possible.*
- ***I hope you'll make this course what you need it to be.*** *We know that learning and creative work cannot happen passively; learning is not something that can be done to or for you—and that's a good thing: the learning is yours. This course is a space where you can actively follow your interests. Find what makes you love your work. Find the motivation and inspiration that will carry you forward. Please talk to me, and I'll be very happy to help you pursue new areas of study. I love suggesting readings or linking you with new resources. I hope we can all do this for each other: swapping resources and sharing what is meaningful to each of us.*
- ***I hope you'll contribute to, and lean upon, our community.*** *Our work is at once collective and individual, and you're invited to pursue forms of self-study while also finding what is fruitful for you within our group. Sometimes we may find that we need to move into smaller groups, or affinity groups, based on what we are working with. Sometimes you may find that you prefer to work in solitude, and other times you may find that there is something special about the meanings and rhythms of moving through a process in collaboration and community. Sometimes the simple act of doing something with someone else makes all the difference. Know that you have choice here, but I hope that we'll come to share resources (and food, if you all agree). I hope we will take care of each other. I hope we will get to know one another as human beings outside of the class. I hope we will celebrate as often as we can.*
- ***I hope you'll show up—because our community relies upon you.*** *Come to our class session/Zoom call each week. We feel your absence when you're not with us. To join in our space is an act of respect for everyone in our community. If you need to be absent from class during one or more weeks, please send our whole group an email to let us know. (No need to offer an explanation, but we will be glad to hear from you.)*
- ***I hope you'll honor our circle.*** *In our time together, I hope we can come around a circle (whether we can physically arrange our desks in a circle or whether we must imagine ourselves in a circle as we appear in our zoom windows). The circle is closed in our brief gatherings, so that we can feel a sense of connection as a group (even as we know we are connected to many other communities within that circle). What is said or written within our space may be in-process, contingent, nascent, or provisional, and it is also offered in a very specific, momentary context. What is shared may be personal and vulnerable to say. Let's not take what is said out of our context. In order to fully trust each other, let's commit to not repeating what colleagues and friends say to others beyond our course community.*
- ***I hope you'll write with your own vision, but also allow the course material to challenge you and push you in new directions.*** *Finding your own way as a writer and*

researcher is a key part of your learning experience in our course, and you may find it at times to be frustrating, challenging, or confusing. Consider this all part of a real creative process—a creative process that I hope will continue for you beyond this course. Writers, researchers, activists figure things out as they go; this course is real-life work. I can't offer a predetermined path; we will find the path together.

- ***I hope you'll visit my office hours at least once this semester.*** *Let's be in conversation. You can make an appointment with me any week of the semester. During our time together, feel free to pose a question or tell me about your semester, your experience of the course, a favorite book you've read, or something you shared with a family member from our course.*
- ***I hope you'll take care of yourself.*** *Your well-being and your education are interconnected; we take both seriously. Make your health a priority. We will come up with a timeline for our shared work so that we can fall into a rhythm together, but you can always let the group know if you need to adjust a due date or timeline; we understand that life is full of the unexpected. Just let us know what you need.*

We want to hear what you hope for this semester, too. We'll take time to discuss, add to, and modify the list you've just read.

Accompanying this letter is another list: a library of readings, exercises, and writing prompts that I have collected along the way, which all speak in some way to what I understand to be our central purpose in this course. I offer this library as a starting point, but we will add it to this list together this semester. I recognize that we will each find different relationships to the texts listed and those we add; reading is a relational practice.

Because it's helpful to move through readings and forms of work together, as a community, we'll select from this list what we want to do and when. We'll revise our plan as we go so that we can be responsive to what is emerging.

You'll always know what to expect from the next week/next class session (we'll use the model of a circular agenda that we'll read about in adrienne maree brown's book Emergent Strategy)*, but our plans may shift as we go. We will balance responsiveness with clarity about expectations.*

I imagine that you might have several questions about what this letter proposes. One question might be: "What about grades?" Let's collectively identify the role grades can play in the vision we have for this course. How can grades honor the unique contributions of each person in our group, rather than reductively quantify work with a scale of letters? Even as we may also have a justified skepticism toward them, how can we use grades to support our self-accountability and accountability to each other? Let's think about this as we also think about what we're here to do, and what we want for and with each other.

Thank you in advance for all you'll contribute to our class.

With anticipation for all the good that's ahead,

Janelle

This letter is just a starting point, one means of beginning the ongoing process of being together. Flexibility and responsiveness are necessary to what follows in a course, while remaining with the commitment to antiracist and anti-oppressive practice. This is a consent-based pedagogy that invites and suggests but does not command. This orientation resists pre-given prescriptions for the classroom, seeking a relational approach to teaching that accounts for the needs, orientations, and positionalities of all who come to the space (both facilitator and students alike), the context and situatedness of each course, and what is possible for each occasion. There is no one-size-fits-all model; an educational experience is co-created, and a community comes together in a dynamic unfolding under a shared sense of purpose.

These suggestions are not intended to neglect the particular knowledges of the environmental humanities, but rather to acknowledge that these knowledges are lent to expert facilitation. We can guide by invitation, suggesting and proposing what will enrich the experience. When we don't proceed in an invitational way, when we lean too heavily on our power, our students may "do what they're told," but we are not meeting them in a place of authenticity. When classroom work is motivated by our authority (e.g., assignments completed because they're assigned), we are working inside of dominant systems rather than in opposition to them. Slowing down enough to find what is truly meaningful to students resists the disciplining machinations of the education industrial complex.

The environmental humanities course can be a means of cultivating what Reverend Deborah Johnson calls "possibility consciousness," the kind of consciousness that has vision enough to pull the world toward transformation. "*Pain pushes* until *vision pulls*," Reverend Johnson writes, describing how we are moved. The cultivation of an inclusive space of belonging is necessary to move toward vision, toward a more just world. As Reverend Johnson notes, "We cannot create what we are not willing to become."[45] Changing the systems we are embedded within reaches to the very core of one's existence in the world.

Notes

1 Rauna Kuokkanen, *Reshaping the University: Responsibility, Indigenous Epistemes, and the Logic of the Gift* (U of British Columbia P, 2007) 72.

2 Eve Tuck and K. Wayne Yang, "Decolonization is Not a Metaphor," *Decolonization: Indigeneity, Education, & Society* 1.1 (2012): 1–40.

3 Tema Okun, "Characteristics of White Supremacy Culture," *dRworks*, www.dismantlingracism.org

4 Carlos Blanco, "Mental Health of College Students and Their Non-college-attending Peers: Results from the National Epidemiologic Study on Alcohol and Related Conditions," *Arch Gen Psychiatry* 65.12 (December 2008): 1429–1437, https://doi.org/10.1001/archpsyc.65.12.1429; Liz Morrish, "Pressure Vessels: The Epidemic of Poor Mental Health Among Higher Education Staff," *HEPI Occasional Paper* 20 (May 2019), www.hepi.ac.uk/; National Alliance on Mental Health, *College Students Speak: A Survey Report on Mental Health*, 2012, www.nami.org

5 Eli Meyerhoff, *Beyond Education: Radical Studying for Another World* (U of Minnesota P, 2019) 62. Meyerhoff historicizes the term "education" through Locke's class-based and racist doctrine that delineates education for the gentry and working schools for poor children. Meyerhoff articulates how this racist and classist orientation still conditions

approaches to education. His theory of "radical studying" sets the practice of liberatory learning apart from the continued violences of educational systems.

6 Sara Ahmed, *The Cultural Politics of Emotion* (Routledge, 2004); Sara Ahmed, *The Promise of Happiness* (Duke UP, 2010); Sara Ahmed, *Willful Subjects* (Duke UP, 2014); Leah Lakshmi Piepzna-Samarasinha, *Care Work: Dreaming Disability Justice* (Arsenal Pulp, 2018). All oppressions are connected; they are not separate and distinct.

7 Meyerhoff observes that educational culture operates with an "affective pedagogical economy of credit and debt—students are disciplined to desire honor and avoid shame in the eyes of their teachers and fellow students, often taking the form of grades on exams." The culture of shame in turn produces what Meyerhoff reads as "binary figures of educational value and waste (e.g., the success versus the failure, the college-bound versus the remedial, the graduate versus the dropout)."

8 Thanks to Reverend Deborah Johnson for this point. Reverend Johnson defines true inclusion as a spiritual attribute that does not go unnoticed because it is transformative. Inclusion means change.

9 Walter D. Mignolo,"Epistemic Disobedience, Independent Thought and Decolonial Freedom," *Theory, Culture, and Society* 26.7–8 (2009): 160.

10 Tara J. Yosso, "Whose Culture Has Capital? A Critical Race Theory Discussion of Community Cultural Wealth," *Race Ethnicity and Education* 8.1 (2005): 69–91.

11 A syllabus-less course allows agreements to be built collaboratively, rather than policies to be assigned. Classroom "policies" are about what a group of people need to agree to in order to promote everyone's mutual thriving. Negotiating these terms of agreement puts teacher and student in relation to each other from the start, charting a path that is mutually beneficial.

12 I use the term "facilitator" rather than instructor or teacher in order to shift assumptions about the role of the person who is compensated for their presence in a classroom space.

13 See Keeanga-Yamahtta Taylor, *How We Get Free: Black Feminism and the Combahee River Collective* (Haymarket, 2017).

14 Leanne Betasamosake Simpson, *As We Have Always Done: Indigenous Freedom through Radical Resistance* (U of Minnesota P, 2017) 9.

15 Simpson, *As We Have Always, Done*, 17.

16 Simpson, *As We Have Always Done*, 17.

17 Neil Roberts, *Freedom as Marronage* (U of Chicago P, 2015).

18 Simpson, *As We Have Always Done*, 19.

19 Simpson, *As We Have Always Done*, 44–45.

20 Simpson, *As We Have Always Done*, 155.

21 Simpson, *As We Have Always Done*, 155.

22 Simpson, *As We Have Always Done*, 43.

23 Patrick Wolfe, "Settler Colonialism and the Elimination of the Native," *Journal of Genocide Research* 8.4 (2006): 388.

24 Linda Tuhiwai Smith, *Decolonizing Methodologies*, 2nd edition (Zed Books, 2012) 175.

25 Macarena Gómez-Barris, *The Extractive Zone: Social Ecologies and Decolonial Perspectives* (Duke UP, 2017) xv.

26 Mariana Mora, *Kuxlejal Politics: Indigenous Autonomy, Race, and Decolonizing Research in Zapatista Communities* (U of Texas P, 2017) 6–7.

27 David Naguib Pellow, *What Is Critical Environmental Justice?* (Polity, 2018) 27.

28 See Stacy Alaimo, *Bodily Natures: Science, Environment, and the Material Self* (Indiana UP, 2010).

29 Dina Gilio-Whitaker, *As Long as Grass Grows: The Indigenous Fight for Environmental Justice, from Colonization to Standing Rock* (Beacon, 2019) 112.

30 adrienne maree brown, *Emergent Strategy: Shaping Change, Changing Worlds* (AK Press, 2017) 41. This idea is adopted from Taj James, from the inaugural meeting of the environmental justice resource redistribution initiative Building Equity and Alignment in 2013.

31 brown, *Emergent Strategy*, 215.
32 brown, *Emergent Strategy*, 42.
33 Priya Parker, *The Art of Gathering: How We Meet and Why It Matters* (Riverhead, 2018) xi.
34 Parker writes of ritual that is rendered in the form and format of a gathering. "People come to expect these elements of form and even take comfort in them. Over time, the form itself plays a role in shaping people's sense of belonging to the group and their identity within that group: *This is who we are. This is the way we do things around here.*"
35 brown, *Emergent Strategy*, 5. See also adrienne maree brown's *Holding Change: The Way of Emergent Strategy Facilitation and Mediation* (AK Press, 2021), published when the present text was in press.
36 brown, *Emergent Strategy*, 143.
37 Kevin K. Kumashiro, "Against Repetition: Addressing Resistance to Anti-Oppressive Change in the Practices of Learning, Teaching, Supervising, and Researching," *Race and Higher Education: Rethinking Pedagogy in Diverse College Classrooms*, edited by Annie Howell and Frank Tuitt (Harvard Educational Review Reprint Series, Number 36, Harvard Education Publication, 2003) 50.
38 adrienne maree brown calls this "the feedback loop—experiment, gather feedback, experiment again. This is how we learn" (*Emergent Strategy* 106). This loop puts strategic work in motion. The principles of coming together in community that are iterative, in-process, and available for revision.
39 brown, *Emergent Strategy*, 220.
40 Beth Berila, *Integrating Mindfulness into Anti-Oppression Pedagogy: Social Justice in Higher Education* (New York: Routledge, 2016) 57.
41 In my own practice, I have found it helpful to share when I am trying something new. I might say: *This is a new way of teaching for me and I'm still feeling my way through. I have been complicit in a system that has produced stress and suffering. With you here, I want to create a community that is enlivening for more of us. I don't have a fixed expectation for what we will be this semester; rather, I have faith in what we can co-create.* I recognize that what I can say to my students is conditioned by my positionality and my context. We can access different forms and levels of vulnerability based on who we are and the privileges we are afforded in different times and places and with different groups.
42 brown, *Emergent Strategy*, 91.
43 Asao B. Inoue, *Antiracist Writing Assessment Ecologies: Teaching and Assessing Writing for a Socially Just Future* (WAC Clearinghouse, 2015).
44 Meyerhoff, *Beyond Education*, 6.
45 Reverend Deborah L. Johnson, "Cultivating the Consciousness of Inclusion," *Education Series*, 2021.

Works cited

Ahmed, Sara. *The Cultural Politics of Emotion*. Routledge, 2004.

Ahmed, Sara. *The Promise of Happiness*. Duke UP, 2010.

Ahmed, Sara. *Willful Subjects*. Duke UP, 2014.

Alaimo, Stacy. *Bodily Natures: Science, Environment, and the Material Self*. Indiana UP, 2010.

Berila, Beth. *Integrating Mindfulness into Anti-Oppression Pedagogy: Social Justice in Higher Education*. Routledge, 2016.

Blanco, Carlos. "Mental Health of College Students and Their Non-College-Attending Peers: Results from the National Epidemiologic Study on Alcohol and Related Conditions." *Archives of General Psychiatry* 65.12 (December 2008): 1429–1437.

brown, adrienne maree. *Emergent Strategy: Shaping Change, Changing Worlds*. AK Press, 2017.

brown, adrienne maree. *We Will Not Cancel Us: And Other Dreams of Transformative Justice*. AK Press, 2020.

Freire, Paolo. *Pedagogy of the Oppressed*. 30th Anniversary ed. Continuum, 2000.
Gómez-Barris, Macarena. *The Extractive Zone: Social Ecologies and Decolonial Perspectives*. Duke UP, 2017.
Inoue, Asao B. *Antiracist Writing Assessment Ecologies: Teaching and Assessing Writing for a Socially Just Future*. WAC Clearinghouse, 2015.
Johnson, Reverend Deborah L. "Cultivating the Consciousness of Inclusion." *Education Series*. 2021.
Kumashiro, Kevin K. "Against Repetition: Addressing Resistance to Anti-Oppressive Change in the Practices of Learning, Teaching, Supervising, and Researching." *Race and Higher Education: Rethinking Pedagogy in Diverse College Classrooms*, edited by Annie Howell and Frank Tuitt. *Harvard Educational Review*, Reprint Series 36 (2003).
Kuokkanen, Rauna. *Reshaping the University: Responsibility, Indigenous Epistemes, and the Logic of the Gift*. U of British Columbia P, 2007.
Meyerhoff, Eli. *Beyond Education: Radical Studying for Another World*. U of Minnesota P, 2019.
Mignolo, Walter D. "Epistemic Disobedience, Independent Thought and Decolonial Freedom." *Theory, Culture, and Society* 26.7–8 (2009): 159–181.
Mora, Mariana. *Kuxlejal Politics: Indigenous Autonomy, Race, and Decolonizing Research in Zapatista Communities*. U of Texas P, 2017.
Morrish, Liz. "Pressure Vessels: The Epidemic of Poor Mental Health Among Higher Education Staff." *HEPI Occasional Paper* 20 (May 2019). www.hepi.ac.uk.
National Alliance on Mental Health. *College Students Speak: A Survey Report on Mental Health*, 2012. www.nami.org
Okun, Tema. "White Supremacy Culture." *DRWorksBook*, 2019. www.dismantlingracism.org/white-supremacy-culture.html
Parker, Priya. *The Art of Gathering: How We Meet and Why It Matters*. Riverhead, 2018.
Piepzna-Samarasinha, Leah Lakshmi. *Care Work: Dreaming Disability Justice*. Arsenal Pulp, 2018.
Roberts, Neil. *Freedom as Marronage*. U of Chicago P, 2015.
Simpson, Leanne Betasamosake. *As We Have Always Done: Indigenous Freedom through Radical Resistance*. U of Minnesota P, 2017.
Smith, Linda Tuhiwai. *Decolonizing Methodologies*. 2nd ed. Zed Books, 2012.
Taylor, Keeanga-Yamahtta. *How We Get Free: Black Feminism and the Combahee River Collective*. Haymarket, 2017.
Tuck, Eve, and K. Wayne Yang. "Decolonization Is Not a Metaphor." *Decolonization: Indigeneity, Education, & Society* 1.1 (2012): 1–40.
Wolfe, Patrick. "Settler Colonialism and the Elimination of the Native." *Journal of Genocide Research* 8.4 (2006): 387–409.
Yosso, Tara J. "Whose Culture Has Capital? A Critical Race Theory Discussion of Community Cultural Wealth." *Race Ethnicity and Education* 8.1 (2005): 69–91.

CODA

No final word

In *Precarious Life*, a book about those lives that are made to become ungrievable under the dehumanization of colonial-capitalist regimes, Judith Butler wagers that "it is possible to appeal to a 'we,' for all of us have some notion of what it is to have lost somebody. Loss has made a tenuous 'we' of us all."[1] It may be possible to consider a "we" affected by the many griefs of climate change and interrelated global phenomena, but that "we" exists with severe, devastating disproportionality.

An epistemically just environmental humanities will mean a fundamental reconstitution of the field in terms of who and what is centered and in terms of knowledge practices, heuristics used in the classroom, approaches to that which we read and write. Simply turning to the matter of ontology is not sufficient for recognizing radical difference in onto-epistemologies of our globe. We need to stay with our field's ability to remake itself as it reveals its own limitations. We are in a moment of remaking.

I have no final word. I know only that I want to stay with the questions posed by McKittrick:

> What is to be done and what are we doing to reimagine how we might live in the world differently? How has this world already been radically reimagined by and through black thought, and how do we share the knowledge with each other so that we might practice a humanism made to the measure of the world?

McKittrick asks us to "collaboratively call into question a system of knowledge that delights in accumulation by dispossession and profits from ecocidal and genocidal practices" and to "produce and share stories that honor modes of humanness that cannot and will not replicate this system."[2]

DOI: 10.4324/9781003037439-7

An environmental humanities that is oriented toward environmental justice will involve the kinds of radical imagination that McKittrick calls us toward. Stories are complemented by a range of strategies: daily practices of healing justice, organizing, policy-making, abolishing extractive economies and state violence, reparations, other forms of material transformation, formalized means of ensuring sovereignty and self-determination of marginalized peoples. These environmental justice strategies cannot be enacted from a single vantage point. They are not bestowed; they are built through coalitional processes that are constantly on the lookout for what and who is being missed in a given situation or moment. These strategies come about through many-worldedness as they also seek to build the world anew, with newness here being also a form of recovery and revitalization as resistance to genocide. To practice environmental justice means seeking a world in which many worlds fit, a world in which self-determination is valued and understood in its fullness. This requires understanding that land is onto-epistemic, that the earth teaches and shapes belief systems, that ways of being are only possible in relation to land, and that existence is fundamentally relational. One of the greatest epistemic violences of colonial systems is the disruption of kinship and connection through the imposition of the belief in land as property, that makes the living into object. Countering these overlapping forms of epistemic violence that are continuous with material violences must be central to environmental justice work and therefore central to the environmental humanities.

Our work as educators is to create spaces where these fundamental concerns are discussed and acted upon in productive ways, to facilitate communities that continue to equip all of us for the stamina and connectedness for social action and change. If the classroom can be made to be a site of coalition-building, mutual aid, and reciprocal learning, we will be leveraging educational spaces more fully. In all of this, our work is embodied. We need practices that enable our fullest participation as we come to know ourselves in other terms than those provided by dominant systems. Insofar as we can be invited into the wisdom traditions that are culturally rooted and, like any form of growth, also continually being remade, we may be able to access ways of thinking/being that call dominant systems into question. We may be more able to nourish a world in which many worlds fit. We may be more able to find and sustain ourselves in such a world.

Notes

1 Judith Butler, *Precarious Life: The Powers of Mourning and Violence* (Verso, 2006) 20.
2 Katherine McKittrick, *Dear Science and Other Stories* (Duke UP, 2020) 74.

Works cited

Butler, Judith. *Precarious Life: The Powers of Mourning and Violence*. Verso, 2006.
McKittrick, Katherine. *Dear Science and Other Stories*. Duke UP, 2020.

INDEX

Note: Page numbers followed by "n" indicate a note on the corresponding page.

Taylor & Francis Group
an informa business

Taylor & Francis eBooks

www.taylorfrancis.com

A single destination for eBooks from Taylor & Francis with increased functionality and an improved user experience to meet the needs of our customers.

90,000+ eBooks of award-winning academic content in Humanities, Social Science, Science, Technology, Engineering, and Medical written by a global network of editors and authors.

TAYLOR & FRANCIS EBOOKS OFFERS:

- A streamlined experience for our library customers
- A single point of discovery for all of our eBook content
- Improved search and discovery of content at both book and chapter level

REQUEST A FREE TRIAL

support@taylorfrancis.com

Routledge
Taylor & Francis Group

CRC Press
Taylor & Francis Group

For Product Safety Concerns and Information please contact our EU representative GPSR@taylorandfrancis.com
Taylor & Francis Verlag GmbH, Kaufingerstraße 24, 80331 München, Germany

www.ingramcontent.com/pod-product-compliance
Ingram Content Group UK Ltd.
Pitfield, Milton Keynes, MK11 3LW, UK
UKHW020002200926
475702UK00015B/2439
* 9 7 8 0 3 6 7 4 7 9 6 2 6 *